# THE ANIMATE AND THE INANIMATE

# THE ANIMATE AND THE INANIMATE

## WILLIAM JAMES SIDIS

ISBN: 978-1-957990-69-9

# Contents

# PREFACE

This work sets forth a theory which is speculative in nature, there being no verifying experiments. It is based on the idea of the reversibility of everything in time; that is, that every type of process has its time-image, a corresponding process which is its exact reverse with respect to time. This accounts for all physical laws but one, namely, the second law of thermodynamics. This law has been found during the nineteenth century to be a source of a great deal of difficulty. The eminent physicist, Clerk-Maxwell, in the middle of the nineteenth century, while giving a proof of that law, admitted that reversals are possible by imagining a "sorting demon" who could sort out the smaller particles, and separate the slower ones from the faster ones. This second law of thermodynamics brought in the idea of energy-level, of unavailable energy (or "entropy" as it was called by Clausius) which was constantly increasing.

In the theory herein set forth, we suppose that reversals of the second law are a regular phenomenon, and identify them with what is generally known as life. This changes the idea of unavailable energy into that of a reserve fund of energy, used only by life, and created by non-living forces.

This is in accordance with some recent discoveries. The late Prof William James has discovered in the domain of mental phenomena what he calls "reserve energy," which later investigation has shown to be present to a more limited extent in all biological phenomena. It remained a mystery, however, where this energy came from, and the theory of reserve energy as set forth in this work suggests a possible explanation of these phenomena.

In relation to the universe as a whole, the theory herein set forth represents the idea of what is known as cyclical change. This idea is a very old one, being found among the philosophers of the Ionian school, and reappearing at later periods from time to time. On the other hand, the generally accepted theory of the second law of thermodynamics represents a different philosophical tendency, the tendency that considers changes once made as irreparable. Aristotle's philosophy is a good example of that tendency in ancient times, but it has appeared more recently, especially in Spencer's theory of evolution, which, it is interesting to note, is hardly more than a statement of the second law of thermodynamics in philosophical terms.

Since the manuscript was completed my attention was attracted by a quotation from a lecture by the great scientist, Lord Kelvin, in which a theory is suggested which is very similar to mine in its general outlines; Lord Kelvin, however, does not work out the theory. He suggests that life works through a reversal of the second law of thermodynamics; and that living organisms, especially animal life, actually act the part of ClerkMaxwell's "sorting demon." Lord Kelvin, however, regards this as an indication of some suspension of the ordinary physical laws, instead of seeking for the explanation of this reversal in these physical laws themselves.

To quote Lord Kelvin's own words: "It is conceivable that animal life might have the attribute of using the heat of surrounding matter, at its natural temperature, as a source of energy for mechanical effect.... The influence of animal or vegetable life on matter is infinitely beyond the range of any scientific enquiry hitherto entered on. Its power of directing the motions of moving particles, in the demonstrated daily miracle of our human free-will, and in the growth of generation after generation of plants from a single seed, are infinitely different from any possible result of the fortuitous concurrence of atoms."

Here the suggestion is obvious that the phenomena of life operate as Clerk-Maxwell's supposed "sorting demon," through reversing the second law of thermodynamics and utilising the unavailable or reserve energy of matter; only Lord Kelvin, instead of deriving this from the ordinary physical laws, immediately concluded that some mysterious vital force must be in operation. Under my theory, this reversal can be explained on the pure basis of the theory of probability.

It is also to be noted that the theory which 1 suggest in this work solves not only the biological problem of reserve energy, but also certain astronomical paradoxes in connection with the theory of the structure of the universe and its evolution.

The latter part of the work, which deals with the theory of the reversibility of time and the psychological aspect of the second law of thermodynamics itself, is a purely speculative section, partaking more of the metaphysical than of the scientific. However, even in that section, it is to be hoped that there will be found a basis for putting the theory of the nature of time on a scientific basis and for taking it finally outside of the domain of metaphysics.

At the end of the work, a number of objections to my theory are stated in order to show what objections can be adduced. 1 do not attempt to answer these arguments, but, for the sake of fairness to the reader, simply state them and leave them unanswered, so that the reader may decide for himself all the pros and cons of the question, and come to a more unbiased conclusion.

At first 1 hesitated to publish my theory of the reversibility of the universe; but 1 was encouraged on discovering the quotation from Lord Kelvin above mentioned; so that now, knowing that this is not the first time that it has been suggested that life is a reversal of the second law of thermodynamics, I have decided to publish the work and give my theory to the world, to be accepted or rejected, as the case may be.

WILLIAM JAMES SIDIS

January 6, 1920.

# I  THE REVERSE UNIVERSE

Among the physical laws it is a general characteristic that there is reversibility in time; that is, should the whole universe trace back the various positions that bodies in it have passed through in a given interval of time, but in the reverse order to that in which these positions actually occurred, then the universe, in this imaginary case, would still obey the same laws.

To test reversibility, we may imagine what we may call "the reverse universe," that is to say, another, an imaginary universe, in which the positions of all bodies at various moments of time are the same as in our real universe, in which those positions occur at the same respective intervals of time but in the reverse order. To assist in imagining this reverse universe, we may remind ourselves that, when we look in a mirror, the imaginary world that we see in that mirror corresponds in every detail to the world we are in, with the exception that one dimension of space occurs in the reverse order, namely the direction perpendicular to the plane of the mirror. If, now, we conceive of time as a sort of additional dimension of the universe, then our "reverse universe" would be one in which there was a similar reversal in that dimension, leaving the three dimensions of space unaltered. Or, to put it in another way, the series of images produced by running a motion-picture reel backwards would give exactly the impression of such a reverse universe.

With this auxiliary, imaginary universe, our test of the reversibility of any given physical law or process would be, whether that law holds good, whether that process still subsists in the reverse universe. In order to see that in any case, we may first find out how to translate any physical occurrence into the corresponding occurrence in our reverse universe. To start with, all positions in space remain absolutely the same in the reverse universe as in the real universe; intervals of time, however, remain the same in magnitude but are reversed in direction. In other words, though the absolute amount of an interval of time remains unchanged, it is necessary, in translating into terms of the reverse universe, to replace "before" by "after," and vice versa.

The path of a moving body will remain the same in the reverse universe because all the positions which constitute that path will remain unchanged. Since, however, the positions are reached in the reverse order of time, the body moves along the path in the reverse direction. The absolute amount of corresponding intervals of space and time in this motion remaining unchanged, it follows that all velocities must, in the reverse universe, be the same in amount but exactly reversed in direction.

9

We come to a problem of greater difficulty in considering what becomes of acceleration. Acceleration is the rate of change of velocity with respect to time. If, to make the question simpler, we assume uniform acceleration, then the acceleration of a body is equal to the difference of velocity divided by the interval of time required to produce this difference. If, for example, in an interval of time T the velocity A is changed to the velocity B, the acceleration (vectorially represented) would be (BA)/T. In the corresponding motion in the reverse universe, in the interval of time T, the velocity changes from -B to A, so that the acceleration is [(-A)-(-B)]/T, or (B-A)/T. In other words, the acceleration of a body remains unchanged in the reverse universe, both in amount and in direction, in translation into terms of the reverse universe. The above reason assumes that the acceleration of the body is uniform, but an extension of the same reasoning will show that the same conclusion holds even when the acceleration is constantly varying.

So much for pure kinematics. For dynamical terms, it is necessary to find what happens to the mass of bodies in the reverse universe. Now, mass being merely amount of matter, and unrelated to time, it follows that mass is not in the least changed by reversal. From that it follows, by what we have seen, that all momenta are reversed in direction but unchanged in amount, while, in the reverse universe, the force acting on a body, being the product of two magnitudes that remain unchanged in the reverse universe (namely, the mass of the body and the acceleration, assuming no other force to act), must necessarily remain unchanged in the reverse universe not only in amount but also in direction. It might have been expected that, in the reverse universe, forces would be reversed in direction; but this is not so.

Energy, being entirely dependent on such things as position and force (in the case of potential energy) or on mass and the square of speed (in the case of kinetic energy), all of which remain entirely unchanged in the reverse universe, must manifestly remain entirely unchanged.

We come, however, to a more complicated problem in the question of the causal relation. For this purpose it is necessary to distinguish various kinds of causality. The true relation of cause and effect is one of temporal sequence; e.g., the removal of the support of an object is the cause of its falling. The force of gravity has been there all the time; and it is a logical consequence of the existence of such force that the fall of an object should follow the removal of its support. Strictly speaking, the force of gravity is in this case not a cause, but an explanation, a reason for the actual causation, which is itself merely a sequence with an explanation. We have thus to distinguish between the relation of reason and consequence, on the one hand, and, on the other hand, the relation of cause and effect. The latter implies sequence in time, the former is a pure relation of logical deduction and essentially implies simultaneity, for the reason and the consequence, one being a logical deduction from the other, must both subsist together. Now, in the reverse universe, we must suppose that all logical relations of facts remain the same. This does not imply anything concerning mental phenomena; of that we shall find out later in our investigation. In fact, logical relations of facts must of necessity subsist apart from the question whether or not a mind exists in the universe. Logical relations may be said to be simply the most general external facts in existence. If A is B and B is C, the rule then is, not that 1 think that A is C; it is a fact verifiable by observation that A is C. Hence, even should the reverse universe destroy completely all mental phenomena, logical relations must remain unchanged, and consequently also the relation of reason and consequence.

But with true physical causality, it is otherwise. If some general law or some particular force resulting therefrom has for its consequence, in the real universe, that event A should be followed by event B, then the corresponding law or, force in the reverse universe must result in the corresponding events A' and B' following one another in the reverse order. That is to say, if one physical event causes another in the real universe, then the event corresponding in the reverse universe to the effect will, in general, cause the event corresponding in the reverse universe to the cause. That is to say, in translating into terms of the reverse universe, "cause" is to be translated by "effect," and vice versa. This, however, is not an accurate rule, there being exceptions, a causal relation being sometimes altogether severed or else unrecognizably altered by the reversal of time.

Again in the reverse universe, such properties as density, specific heat, elasticity, amount of heat, temperature, etc., also remain unchanged. It could also be shown that such properties as electricity and magnetism remain unchanged, but that the direction of an electric current would be reversed. Thus all physical phenomena could readily be translated into terms of the reverse universe. The various varieties of substance, depending on the internal structure of the atom and molecule, etc., also remain unchanged in the reverse universe.

## II REVERSIBLE LAWS

Now we shall attempt to find out what are the physical laws which subsist in this imaginary "reverse universe." To start with the simple laws of mechanics, we have it given in the real universe that a body retains its velocity unless there is some external force to change that velocity. Now, as there can be no change of velocity in the reverse universe without a corresponding change of the reverse velocity in the real universe, and since all forces in both universes are respectively equal, it follows that this same law of motion applies also in the reverse universe. In other words, the law of inertia is unchanged by the reversal in time, and is therefore what we may call a reversible physical law.

The second law of motion is that change of momentum is proportional to force impressed. Now, following the reasoning which we have already followed in the case of accelerations, the rate of change of momentum remains unchanged in our reverse universe. Furthermore, we have already seen that, in the reverse universe, the force impressed on a body remains unchanged. Hence it follows that the second law of motion subsists in the reverse universe, and is therefore reversible.

The third law of motion is that to every action (force) there is also an equal and opposite reaction. This law is also obviously reversible, since in the reverse universe neither the magnitude nor the direction of forces is altered.

Energy being the same in the reverse as in the real universe, it similarly follows that the law of the conservation of energy holds in the reverse universe, and is therefore reversible in time. The same holds true of the law of the conservation of matter.

One of the principal methods by which motion can be changed is by impact. An impact may be elastic or inelastic. In the case of the very smallest particles of matter, that kind of collision only is possible which loses no energy, but in which the kinetic energy remains the same as before, that is to say, an elastic collision; for, in the case of ultimate particles, none of the original energy can be changed into internal motion of the particles. Hence only elastic collision is possible in the case of ultimate particles; and it is not difficult to show that, in the case of elastic collision, the reversed final velocities of the same masses will cause, as an effect of the impact, the reversed initial velocities. Now, since all matter is made up of these particles, whatever they may be, and therefore all collisions of bodies of matter are made up of countless elastic collisions of ultimate particles, it follows that, in the reverse universe, where impact occurs, all particles of matter follow the same course as they would in the real universe under the same initial conditions. Hence the laws of impact, when brought down to ultimate particles, are perfectly reversible and also remain unaltered in the reverse universe.

The various laws of attraction and repulsion that are found to subsist in the objective universe, such as gravitation, electrical and magnetic attraction and repulsion, etc., dealing as they do with the directions of forces, must also remain unchanged in the reverse universe. Similarly with many other general physical laws.

Even the laws of reflection and refraction of light will remain unaltered in the reverse universe, and are therefore perfectly reversible.

As a result, we may say in general that, looking to the ultimate particles of matter, enough physical laws subsist in the reverse universe to determine, from the positions and velocities of all particles of matter at a given instant, the entire past and future of the universe. The result is that, given those physical laws which we assume to remain always true, if we should imagine that, in the real universe, at one given moment, all particles of matter should, while retaining their respective positions, reverse their velocities, it would follow that this would be enough, of itself, to make all particles of matter trace back their previous positions in the reverse order and thus, as it were, create a reverse universe.

## III IRREVERSIBILITY

So far, we have seen that the physical laws essential to the determination of the course of the universe from its present momentary condition are all reversible. From this it might be concluded that all physical laws must in consequence be reversible, and that, therefore, there can be no essential difference between the real universe and the reverse universe. And this much is true, that, provided we examine the motions of the particles of matter, everything that happens in the reverse universe can be described in terms of the physical properties of matter as we know them.

But at the same time, if we take the most ordinary events of the real universe and attempt to find out what is the corresponding event in the reverse universe, something strange will at once impress us about the reverse universe. Take this, for example: a ball rolls down a staircase, bounces a little at the bottom, and finally stops. In the reverse universe the initial condition is the ball at the bottom, on a floor near the foot of a staircase. The heat energy in the floor collects at one point underneath the ball, so as to push the ball suddenly upward. Each time that the ball falls back to the floor this process is repeated, until finally the floor throws the ball on to the first stair. The stairs, each in turn, throw the ball in a similar manner up the staircase, till finally the ball stops at the top. The molecular vibrations in the ball, floor, and staircase, had previously been so arranged that concentration of energy would happen at a particular spot and time, while the ball so moved that it just happened to be at those spots exactly in time.

So it will be with the occurrences corresponding in the reverse universe to almost any common occurrence in the physical world of our experience. Everything seems to be perfectly explicable in terms of physical laws, but at the same time the combinations of motions seem to have something utterly strange about them. Hence there is some point of difference between the real universe and the reverse universe, and hence there must be some property of the real universe that is irreversible.

This irreversible property is found in what is called the second law of thermodynamics. This, taken in its most general aspect, amounts to this: that the energy of the universe is constantly running down to one co mm on level. In other words, where energy of the same variety is present in different degrees of concentration, those differences will be equalised, and energy of a still higher level or to a greater amount must become dissipated in order to recreate these differences of concentration. Of the various varieties of energy, all kinds tend to turn into heat, which is the least concentrated form of energy; and, even though some of that heat may be re-converted into some other form of energy, still, at each step, some energy is irretrievably lost in the form of heat.

This physical law, as well as all those which are derived from it, is irreversible. Furthermore, only such physical laws as are derived from the second law of thermodynamics are irreversible; so that this law constitutes the sole difference between the real and the reverse universe. Where, in the real universe, energy runs down to a common level, it follows that, in the reverse universe, energy tends to build itself up into different levels.

We may say, then, that the characteristic irreversible part of the universe consists in this, that energy tends to evolve (or devolve) from molar motion of extremely large masses, which is the most concentrated form of energy, to a condition in which all energy is in the form of heat, which is the least concentrated form, and at a uniform concentration, that is to say, at a constant temperature throughout. A final condition would result in which a dead level of energy would be reached, and after that nothing further could ever happen in the universe.

The fact, for instance, that perfectly elastic collisions of large masses of matter do not occur, but that such collisions are inelastic, is a direct consequence of the second law of thermodynamics. The characteristic of an inelastic collision is that some of the molar kinetic energy of the colliding bodies is lost by the impact. This lost kinetic energy is changed into heat, which is always produced by an inelastic collision. This is in strict accord with the second law of thermodynamics. In the reverse universe, on the contrary, an impact would be an occasion for heat to be converted into molar motion, thus increasing the total amount of kinetic energy. Such a collision we may call super-elastic, and is not within our experience.

Again, the resistance offered by one body to another, whether in the form of friction or otherwise, is but an example of the second law of thermodynamics, being another case of change of molar energy into heat. In the reverse universe, the very opposite process would take place. Accordingly we find as might be expected, that the laws of friction, etc., are irreversible.

Many chemical reactions are irreversible, though some are reversible. As a general rule, the irreversible chemical reactions are cases of conversion of chemical energy into heat, in accordance with the second law of thermodynamics. So with all irreversible processes.

In the case of a machine, the ratio of the energy obtained to the energy put in (usually expressed as a percentage) is called the mechanical efficiency of that machine. The remaining energy, that the machine has lost, becomes heat. The second law of thermodynamics, expressed in terms of mechanical efficiency, means that all physical phenomena have a mechanical efficiency of less than 100%. The reverse universe, on the contrary, is distinguished from the universe of our experience in that the mechanical efficiency of its phenomena is over 100%.

Again, to express it in another way. Suppose two bodies, one at a temperature of 0° Fahrenheit, the other at a temperature of 200°. The only available heat-energy in those bodies would be the amount represented by 200 degrees in the hotter body. At the same time, the colder body being 460 degrees above absolute zero, there is unavailable energy, which, according to the second law of thermodynamics, cannot be reached, amounting to 460 degrees in each of the two bodies. If both bodies have the same mass and specific heat, the energy which, under the second law of thermodynamics, is available for conversion into other forms of energy, could thus be represented by 200, while the total heat-energy in the two bodies would be represented by 460+660 =1120. The ratio of available to total energy in this case would be 200:1120, or 5:28. In other words, only 18% of the total heat-energy is available for conversion. The second law of thermodynamics states, not merely that not all the available energy can actually be used for any purpose except heat, but also that all energy in an available form (a form other than heat, or else heatenergy in the form of a difference of temperature) tends to turn into unavailable energy, that the amount of available energy in the universe is constantly decreasing.

In the reverse universe we have a different situation, since the second law of thermodynamics is irreversible. Even the heat-energy below the temperature of the coldest bodies in the environment is not merely available, but constantly drawn on. The same i mm ense fund of energy which in the real physical universe is constantly stored up and unavailable, now ceases to be unavailable, but becomes a reserve fund of energy with which difference of concentration of energy is constantly being built up. Under the second law of thermodynamics a reserve fund of energy is constantly stored up in the form of heat and never afterwards touched; under the reverse of that second law, on the contrary, we start with this reserve fund of energy and constantly draw on it to build up energy differences.

## IV THE PARADOX

The second law of thermodynamics is, as we have seen, an irreversible physical law, and seems to be the one distinguishing characteristic between the real universe and the reverse universe. At the same time, that law is of such a nature, that, for the ultimate particles of matter, it does not exist; it is essentially a law concerning transformations of energy of large masses. And yet all large bodies are made up of countless numbers of the ultimate particles of matter, the laws of whose motion are all perfectly reversible. All phenomena of the reverse universe, however strange they may look, are perfectly explicable in terms of the ordinary physical laws as applied to the smallest material particles. It would seem, then, as though there must be some reason in terms of the reversible physical laws why the second law of thermodynamics must be true; that is, the second law of thermodynamics, if true, should be a consequence of the reversible physical laws applicable to ultimate particles. We are, then, confronted with the paradox of having to deduce an irreversible law from perfectly reversible ones.

And yet, since the reverse universe consists of a perfectly consistent series of positions, obeying all reversible physical laws, it follows that any logical deduction from premises which are reversible laws must inevitably apply to the reverse universe, and that therefore the conclusion must be true in the reverse universe as well as in the real physical universe. That is to say, any deductive conclusion from reversible laws must itself be reversible. And yet, in the case of the second law of thermodynamics, the reversible laws which govern the motions of ultimate particles of matter seem to compound themselves somehow into the best possible example of an irreversible law governing the motions of large masses.

We are, therefore, inevitably led to the conclusion that the second law of thermodynamics cannot be deduced from the reversible laws by strict deductive reasoning. The reversible laws must of necessity leave some room for the possibility of the truth of the reverse of the second law of thermodynamics. But, since the second law of thermodynamics simply represents a general tendency, we come to the conclusion that the only possibility that the second law of thermodynamics represents a correct physical law, is, that it is to be deduced from the reversible laws not as a strict logical consequence, but as a great, or even an overwhelming probability. Such a solution of this paradox of the second law was propounded by ClerkMaxwell and other physicists of the middle of the nineteenth century.

Let us, then, examine the reasoning by which ClerkMaxwell was enabled to reconcile reversible premises with an irreversible conclusion. According to his reasoning, both processes are physically possible, concentration and diffusion of energy. The one process obeys the second law of thermodynamics, the other reverses it. Under the second law of thermodynamics, a collision of large masses will generate heat (conversion of molar energy into heatenergy); under its reversal, the heat generates molar motion in and of itself Now, says Clerk-Maxwell, if particles move in a group, or rather in two approaching groups, the particles are likely to strike one another at all sorts of angles, so that, after the impact, the resulting velocities will become scattered, which means that some of the energy will be converted into heat. On the contrary, a reversal of the process means a concentration of the motions of the particles at the very point and time of the impact, which is a very much more improbable combination, and, requiring as it does that this concentration should happen in a particular direction, at a particular point, at a particular time, in order to have the desired effect, it follows that such a reversal of the second law of thermodynamics is so overwhelmingly improbable as to be almost impossible. The second law of thermodynamics is thus based not on necessity but on extreme probability. A reversal of the second law is possible under the reversible physical laws, as we have seen, but this reasoning tends to prove that it is overwhelmingly improbable, and therefore would almost never happen.

But, again, if the premises of the reasoning are, as we suppose, reversible physical laws, it must be possible to apply the same reasoning to the reverse universe. Consequently, a similar line of reasoning, which must be exactly as correct logically, can be followed by tracing events backwards from effect to cause instead of tracing from cause to effect, as Clerk-Maxwell has done.

Any momentary condition, of the universe may be regarded either as the cause of all future conditions of the universe or as the effect of all past conditions. And not only can a given momentary condition of all particles in the universe determine one and only one possible effect, one and only one possible future; that same given momentary condition (position and velocity of every particle) could only have been caused by one possible past series of conditions. Hence it is just as possible to trace our causal relations step by step backwards, as it is to trace them similarly forwards.

Now, tracing causation thus backwards, we find that molar motions, when traced backwards into the past, will, in all probability, bring us to a time when two masses which are now in motion have been together, in contact. Following Clerk-Maxwell's reasoning, we must say that, when two particles move away from contact with each other, an impact must have been the cause, at least some form of impact of particles, but it is a form of impact which produced molar motion. In all probability, those two particular masses will not have motions which trace back to a rebound of all particles at the same angle; which necessitates, according to the rules of elastic collision, that before the impact the motions of the particles must have been scattered. Thus, tracing the reasoning backwards, we arrive at the probability that the molar motions must have been partially at least caused by heat, that is, to the probability of a reversal of the second law of thermodynamics. On the contrary, in order to have a case in accordance with the second law of thermodynamics, on this analogous reasoning, it would be necessary to suppose two bodies being traced back to contact at some particular time, and that the heat-motions of those bodies, when thus traced back, should suddenly, at the particular moment and point of contact, trace back to a concentration of motion of the particles of each body away from the other, for only such concentration could be the effect of a molar motion bringing the bodies into collision. Now, the probability of such a combination is extremely small, so that, by merely shifting our reasoning gear into reverse, the very same reasoning tells us that the second law of thermodynamics is most extremely improbable, but that, on the contrary, its reversal is an overwhelming probability.

Tracing thus from a given momentary condition of the universe, our forward and backward reasoning combined might be interpreted, if such reasoning could be trusted, to mean that the second law of thermodynamics holds good as a probability as to the future, but that its reversal holds true as to the past. Aside from this result being untrue in point of fact, it is self-contradictory, for any given moment of time is always future as to moments that precede it, and past as to moments that follow it. It follows, then, that there must be some fallacy in Clerk-Maxwell's reasoning, which, when extended, gives us the second law of thermodynamics in the general form.

Take the special case that we have been using as an illustration. Molar motion without heat, it is true, is likely, as a matter of pure theory, to produce, after impact, less molar motion and some heat (the total amount of energy remaining invariable). But such an initial condition is, in itself, extremely improbable. If initial velocities of particles may be selected initially as in any direction: and in any amount, it is extremely improbable that all the velocities will have the same direction and amount, or even approximately so. The smaller the number of particles, the greater the probability of a concentrated motion resulting. Also, the smaller the mass, the greater the probable average velocity of the mass at a given time, when the particles are moving at random. Hence, when there is impact of bodies in which particles move at random, the probabilities are that, at that moment, at the point of contact, the small mass of particles in the immediate vicinity will have a greater speed in all probability than the entire mass. Thus, when the collision occurs, the force available for producing molar motion will consist, in the immediate vicinity of the point of contact, of two average speeds greater than those of the respective masses. If those greater speeds lend to be more towards one another than the masses as a whole, then it would be most probable that some of the heat-energy of the two bodies will be converted into molar motion. On the other hand, if the respective speeds in the vicinity of the point of contact are more away from each other than the velocities of the masses themselves, the reverse will happen. Besides, while we have this possibility of heat turning into molar energy or into some other form of energy, and of differences of energy concentration building themselves up in this manner, we have the contrary tendency supplied by Clerk-Maxwell's reasoning. The result is, that we as yet can form no conclusions as to which tendency is more likely.

If, furthermore, we consider that we must regard for a given moment of time, all positions and velocities as equally likely, and that for all such initial positions and velocities which will give a universe obeying the second law of thermodynamics, there is a reverse universe, equally probable; reversing that law, we come to the conclusion that the second law and its reverse are equally probable. If this is true for any given event, then the probability of the observed facts, that is to say, that all events obey the second law, must be infinitesimally small. So that, again, we are forced to the conclusion that the second law of thermodynamics, being an observed fact which can only be explained as an extremely probable result of the reversible physical laws is, on the contrary, most extremely improbable.

Not merely that, but the second law of thermodynamics, when pushed to its logical conclusion, produces rather absurd results. In the first place, we have, seen that it involves a sort of death of the universe in the remote future, a time when all will be one dead level of heat; though all this will, in all probability, come about slowly. But the rate of decrease of the available energy under this second law is approximately proportional to the amount of available energy in the universe; therefore the rate of the running down of energy into the unavailable form must be constantly decreasing. Tracing backwards, we find that, in the past, the farther back we go, the more we get a larger percentage of available energy in the universe, increasing at an ever greater rate. Therefore it follows that we must arrive at some definite time in the past—and that not at an infinite time back—when the available energy was 100% of the total energy of the universe. At a time probably not much farther back, all the motion in the universe must have consisted of molar motion of masses which, as we go back, must have increased in size till we arrive at a time when all the energy must have consisted of the energy of two halves of the universe moving together, each half of the universe being at a temperature of absolute zero and all its parts moving side by side at exactly the same velocity. This possibility, it is true, is somewhat corroborated by the fact that at present the stars are moving in two opposite directions, in two opposite currents, as it were, which may be supposed to be the remnants of the two original large groups of stars whose collision formed the present universe according to this hypothesis.

At the same time the two original halves of the universe cannot have been altogether mutually impenetrable, for in that case the result of the collision would but have made them rebound, though producing a great amount of internal heat-energy in each, and possibly breaking some small pieces off each. It would seem, then, as though the original halves of the universe must have consisted of separate dark stars, with a structure somewhat similar to the present universe. At the time of the collision, all the stars, even all the particles, in each semi-universe must all be moving together at the same speed and in the same direction.

The second law of thermodynamics, then, must date from some sort of Great Collision out of which the present universe evolved. But what happened before this Great Collision? The answer would have to be, everything was at a temperature of absolute zero, there were two semi universes which were moving towards each other, in each of which there was not even a trace of relative motion. Although each of the two semi-universes was in motion, yet within each there was no motion, no internal energy.

But if such was the situation at the time of the Great Collision, it cannot have been so for an eternity past, unless we conceive of the law of gravitational attraction not to have been true in those times. Taking each semi universe by itself, its reverse universe will also show the same conditions as we have already described, except that the semi-universes are moving away from each other, so that we can proceed in peace without danger from the impending Great Collision. Each semi-universe may, for the purpose of internal occurrences, be regarded as at rest. Gravitation will then draw all the stars of each semiuniverse towards its center of gravity, till all of them fall in there. Reversing once more, so as to obtain the process as it must have been supposed to happen, we get the following result: Each semi-universe originally consisted of one great body; suddenly, somehow, that body exploded into pieces, which formed stars, each piece, though, remaining at a temperature of absolute zero. Finally, in each semi-universe, mutual gravitation of the stars slowed them down to relative rest. Just when this relative rest was reached, the two semi-universes collided, and out of this collision came our present universe. Thus we trace a little farther back to the Great Explosions; but these explosions cannot possibly be traced back any farther according to the known physical laws without violating the second law of thermodynamics. In consequence, if we wish to preserve the second law of thermodynamics, we must either dispense with some of the other physical laws, or as some physicists have done, intersperse a creation. In other words, the second law of thermodynamics cannot have been true for an eternity past, though it may be true for on eternity in the future. And even the assumption of a creation would be assuming a process different from the processes coming under the ordinary physical laws.

In other words, we come to the inevitable conclusion that the subsistence of the irreversible second law of thermodynamics in the same universe as the reversible laws concerning the motion of particles is a paradox, both from that point of view and from the fact that this second law, pushed to its logical conclusion, leads back to a mysterious creation which denies all physical laws whatever.

## V  THE PROBABILITIES IN THE PROBLEM

To help us towards a solution of this paradox, we must first find out what the probabilities actually lead us to conclude. We have already seen that, in a given case, the chances are even as to whether energy will run down or build up. There are also small chances of a neutral condition, in which energy remains, on the whole, at the same difference of concentration as before. But the probability of this neutrality is negligible, and we may say that the probabilities are, that in 50% of the cases the second law of thermodynamics will be obeyed, and in 50% of the cases it will be reversed. If such is the case, the universe as a whole will be neutral; that is, taking all the occurrences over all of time and space, there will be no tendency in one direction or the other.

In this reasoning we can be assured as to the probabilities in any given occasion, for we must assume all combinations of initial positions and velocities to be equally likely. Inasmuch as any event occupies a certain amount of time, let us figure on the probabilities of the positions and initial velocities at the middle of that interval. For any range of positions and velocities resulting in a combination obeying the second law of thermodynamics, we have an equal and therefore an equally probable range of positions and velocities reversing that law; namely, the identical positions with the reverse velocities. Where the positions and velocities happen to border between the two kinds of combinations, we will have a sort of neutral result, which is so improbable as to have a zero probability (though that does not make it impossible). Aside from that, the second law of thermodynamics is, on any occasion, equally probable with its reverse, and the probability of each may be taken as 50%. The probability of the second law of thermodynamics being followed on two certain occasions is, as a result, only 25%; and so on, while its probability for all occasions is almost a nullity.

The probability is, however, as a result of this 50% probability, that approximately half the events of the universe, taking all of space and time, will be in accordance with the second law of thermodynamics, while about half will tend to reverse it. The former tendency we will, for short, call the positive, while the latter tendency we will call the negative tendency.

Between these two there is a bordering, or neutral, tendency, which, as a whole, neither builds energy up nor levels it down.

The universe as a whole, including all of time and space, will tend towards this neutral tendency, but this neutral tendency will simply be a compound of positive and negative tendencies at different parts of space and time tending to cancel one another. Taking definite portions of space and time, the chances are that there will be some sort of preponderance of tendency in one direction or the other, the preponderance being greater the smaller the section of space and time that we take into consideration. We may, therefore, assume that, in the part of space and time under our observation (which, we know is very limited) the preponderance is towards the positive tendency. We may suppose that there are other parts of spaee, and other periods of time, when the preponderanee will be in the reverse direction.

But even where the preponderance is toward the positive tendency, it still remains merely a preponderance, and instances of the negative tendency would be almost certain to occur. It is true that the probabilities are that, in such a part of space and time, instances of the negative tendency will occur to a very limited amount; but, all the same, they will occur.

The probabilities of the situation, then, are as follows: the whole universe, including all of space and time, will tend to have as much of the positive as of the negative tendency, with a certain amount of the neutral tendency. At a particular moment of time the probabilities are that there will also be about as much of one tendency as of the other, but that in some sections of space there will be a preponderance towards the positive tendency, while in other sections of space the preponderance will be the reverse; about half of space falling under one heading, and about half of space falling under the other. In each of those portions of space there will be instances of events opposed to the prevailing tendency, presumably in certain material objects. The same applies to a more limited extent if we take one section of space with respect to the different moments of time.

## VI  SOLUTION OF THE PARADOX

We have seen that the second law of thermodynamics, if pushed to its logical conclusion, leads to absurdities; that, on the basis of the other physical laws, it is most extremely improbable; and that it cannot have been universal for all time past unless we assume some sort of creation or some other form of miracle. On the contrary, we have seen that the probabilities from the physical laws governing the motion of particles, which are all reversible, and whose consequences must therefore be also reversible, lead us to the conclusion that, though the universe as a whole will tend to be neutral in that respect, yet, in certain limited portions of space and time, the second law of thermodynamics represents a prevailing tendency. We may easily, therefore, suppose that the portion of space and time under our observation (which, as we know, is very limited) is just such a section, and that the second law of thermodynamics represents a prevailing tendency of energy to level down in our vicinity and in our epoch. This would seem to be the only way leading out of the paradox which seems to follow from the second law of thermodynamics; so that as this law is thus supposed to be true only for a limited epoch, there is no necessity to suppose any creation or other miracles; and therefore the rule for the whole universe is really reversible.

This would apparently solve our paradox, if not for the fact that, according to this proposed solution, the second law of thermodynamics would represent, not a constant law, as observations would indicate, but, on the contrary, merely a prevailing tendency, with a number of instances of reversals of that law in our own part of space and time. Thus we find a difficulty in accepting this solution of the paradox, namely, that our proposed solution requires that, even in our own section of space and time, there must be many instances of the reversal of the second law of thermodynamics; which seems contrary to observed facts.

And yet, considering that the second law of thermodynamics itself leads to absurdities, it might be worth while to inquire whether, after all, there might not be in our portion of space and time certain instances of the reversal of the second law, certain events with what we have called a "negative tendency," which might have escaped our attention.

In order to conduct this inquiry, we would have to find some way to recognize such a reversal, should any be found. This can be done in two ways: either by translating, co mm on occurrences into the reverse universe, and thus familiarising ourselves with how such a reversal looks (a moving-picture outfit could easily bring this reverse universe before the sense of sight, by operating the reel backwards); or else we can reason from the abstract second law itself and infer from its reversal certain easily recognisable outstanding characteristics. We shall proceed in both ways, starting with the abstract method, then using the other method to fill in, as it were, by way of illustration.

One characteristic of the second law of thermodynamics is that there is, under it, a tendency that large causes should produce smaller effects (some energy becoming lost always in spreading heat throughout the universe), while small causes rarely, though occasionally, produce large effects. Now, since it is always possible to regard any event either as caused by past conditions (reasoning from cause to effect) or as being the cause which will produce the conditions of the future (reasoning from effect to cause), both cause and effect of a given event being a determined thing, we may say that, under the second law of thermodynamics, since a given event is likely to have more visible causes and less visible effects that itself, it follows that, under the second law of thermodynamics, it is easier to explain an event as the effect of past causes than as the cause of future effects. In other words, under the second law of thermodynamics, though reasoning from effect to cause is possible, it is almost necessary to reason from cause to effect, as the physical sciences usually do.

On the contrary, when we have the negative tendency, when the second law of thermodynamics is reversed, the reverse is the case. Under the negative tendency, energy is constantly being reclaimed from the enormous heat reserve which otherwise lies unused, and this will be happening at every occurrence taking place under the reversed second law. Thus the tendency in such a case will be that, while occasionally large causes will produce smaller effects, yet as a general rule smaller causes will produce larger effects. In other words, a given event is most likely to have less visible causes and more visible effects than itself, so that, if we try to explain an event as the effect of past conditions, we shall always have difficulty, because part of the cause in any case, and sometimes even the entire cause, will consist merely of diffused and undifferentiated energy which cannot be observed unless we can keep track of every individual particle of matter. But, on the contrary, if we try to explain such an event as being the cause determined by future conditions which are its effects, such an explanation is simple, because the full effect is observable, and the effect is usually more visible than the cause.

The result is that we get one distinguishing characteristic of that reversal of the second law of thermodynamics for which we are looking. If we find such a reversal, we will, in all probability, be finding some sort of events which it is easier to explain from the future than from the past; in other words, we must, in looking for such a reversal, look for something which, while it acts under the ordinary form of causation like the common physical bodies, yet appears teleological in nature. This teleology is only apparent, for causation under the negative tendency is no different from ordinary physical causation. In causation in general, the reverse or pseudo-teleological explanation is always possible, but is more obvious in the case of a reversal than in the ordinary case of positive tendency. Thus, when we wish to find a reversal of the second law of thermodynamics in our section of space and time, we must look for phenomena with an appearance of teleology.

Another outstanding characteristic of a reversal of the second law is the ability to use the immense store of energy which, under the second law of thermodynamics, is unavailable. In other words, a reversal, besides the property of apparent teleology, must also possess the property of ability to use a store of reserve energy, some of which is always used, while at times even all of it could theoretically be used and converted into visible forms.

So we thus get theoretically two outstanding characteristics of the reversal for which we are looking; namely, apparent teleology and the ability to use a fund of reserve energy. If we can find anything in our section of space and time which has these two properties, then in all probability we have found the reversal for which we are looking.

Now, to take the more concrete method, that of observing the reverse universe, either by reversing any co mm on occurrence, or else in observation by reversing a motion-picture-film, etc. We have already seen that a reversal of such an incident as a ball rolling down a flight of stairs becomes, in the reverse universe, the following: the floor and the stairs successively throw the ball upstairs; the ball itself aids the process by giving a jump, as it were, each time it lands. This would give floor, stairs, and ball somewhat an appearance of being alive. In fact, in any case, all ordinary physical objects will act in the reverse universe somewhat as if alive. Instead of rivers running down to sea, we would have in the reverse universe the situation of seawater rejecting its salt and then jumping up the river channel to the source, where the water, separating itself first into drops and then finally into molecules, makes a final jump up to the clouds; in other words, the water is constantly jumping upwards, as though of its own violation, and aided at each step by the ground pushing it upwards or even throwing it up. Here again there is an appearance of life in objects that we would certainly, in our universe, consider as dead.

Take a more complicated instance: The behavior of drops of mercury on a smooth surface, consisting, we may suppose, partly of metal. These drops, in our universe, would roll around under the influence of any external forces that may happen to be present, unite if two happen to come together, and, in case they touch metal, the drop will shrink and partly amalgamate with the metal. In the reverse universe, on the contrary, we have a different arrangement: the drops will roll around as before, but, in their rolling, will avoid the pure metal surfaces, but will tend to roll over the amalgam surfaces. When in contact with the amalgam, they will extract the mercury, and thus the drops will keep growing. When the drop grows in this manner to a large size, there will appear a constriction, and finally a division into two drops, each like the original. This action of ordinary mercury drops in the reverse universe corresponds in many details to the growth and division of living cells in our universe.

In short, we may say that, in general, events in the reverse universe appear as though they were living phenomena; and the general events of the reverse universe may be taken as the type of negative phenomena, of the reversal of the second law of thermodynamics. We should thus expect, in the real universe, to find such reversals in some sort of living or apparently living phenomena. Furthermore, if we find in the reverse universe some phenomena that, contrary to what might be expected, obey the second law of thermodynamics, it must follow that the corresponding phenomena in the real universe must be precisely those reversals for which we are looking.

Inasmuch as we have seen that ordinary inanimate phenomena take on an appearance of life in the reverse universe, let us see what becomes of living phenomena in the reverse universe. Let us inject some sort of living agency into any previous illustration of the reverse universe. Suppose, in the case of the ball rolling downstairs, that it was originally thrown by someone. The beginning of the incident (which will correspond to the end in the reverse universe) will consist of a human arm starting to move, carrying the ball forward against the resistance of the air, finally letting the ball go, after which the ball, on the momentum thus acquired, proceeds to bounce down the stairs. In the reverse universe the ball properly aided by the floor and the stairs, comes jumping upstairs into the hand; the ball, though it tends to be speeded up by the air pushing the ball along, and by the heat-energy of the ball similarly reacting on the air, yet slows down quickly and finally comes to a stop: the acquired momentum of the ball moves the hand, swings the arm, and finally the molar energy thus transferred to the arm becomes transformed into heat, and the arm stops. This very last part of the incident is a rather unexpected case of the second law of thermodynamics in the reverse universe; and we may note, as a result, that the living body, when reversed, becomes a mere obstacle instead of a moving force. We may therefore conclude: first, that inanimate phenomena, when reversed, become animate: second, that animate phenomena, when reversed, lose the appearance of animation; and third, that animate phenomena, when reversed, lose this appearance because, when reversed, they tend to follow the second law of thermodynamics. The logical conclusion from these would be: that inanimate phenomena are positive tendencies, and follow the second law of thermodynamics, while animate phenomena, on the contrary, are negative tendencies and tend to reverse that law. Thus we have found where our part of the universe contains reversals, and come to a solution of our paradox

## VII THEORIES OF FIFE

We find that the theories of the nature of life divide themselves into two varieties: the mechanistic and the vitalistic. The former kind of theory states that all living phenomena are to be explained solely by the ordinary physical laws, and that life differs from other phenomena only on account of its complexity, or in some other incidental manner. On the other hand, the vitalistic theories are to the effect that living phenomena are characterized by some mysterious sort of "vital force" which would seem to have the power to suspend or alter the operation of the physical laws that govern the rest of the universe. In the course of the history of science, much has been said both for the vitalistic and the mechanistic theories, and, as yet, no agreement has actually been reached on that subject.

In the attempt to solve our paradox of the second law of thermodynamics, we have incidentally reached a suggestion of the nature of life. According to the conclusions we have reached, there are in the universe what we have called positive tendencies, neutral tendencies, and negative tendencies, all of which are possible results of the reversible physical laws governing the motion of particles of matter. The neutral tendency being an extremely improbable result, very few cases of it are likely to take place; but, in any given case, unless further special circumstances alter the probabilities, the positive or the negative tendency has a 50% probability, and will therefore result from the reversible laws in about half of the cases occurring in the universe. In our section of the universe the positive tendency, however, preponderates, though, inasmuch as it would be extremely improbable that any section of the universe is entirely without instances of the negative tendency, it follows that there must be phenomena of the negative tendency within our observation. The phenomena of the negative tendency are the living phenomena; while the phenomena of the positive tendency are the non-living phenomena.

This theory of life is strictly mechanistic in so far as life is assumed to operate solely under the physical laws applying to the motion of particles, which laws are sufficient to determine a complete chain of causation. On the contrary, physicists, confining their observation entirely to inanimate matter, have reached the conclusion that there is a further physical law, the so-called second law of thermodynamics, which is suspended by living phenomena. There is according to our theory, this essential difference between living and non-living phenomena; and this difference would supply the basis for the idea of "vital force." Thus the two theories of life can be reconciled.

On the matter of the difference between living and nonliving bodies, there is still less agreement. For instance, it is stated that lifeless substances, in so far as they form definite shapes, form only geometrical shapes, while living substances form irregular shapes. Outside of the fact that this does not distinguish living bodies from bodies which were once alive but which have lost the property of life, and outside of the fact that not all inorganic substances but only certain solid substances form geometrically shaped crystals, we may refute the statement that living bodies always have irregular shapes by simply adducing the example of the egg. This distinction is therefore on all sides untenable.

Again, it has been said that the difference between living and lifeless substances is the question of the presence of organs. But will that alone distinguish the average organism from a machine? The same objection can be urged against the proposed distinction on the ground that living bodies have a complex organization. However, either of these proposed distinctions may mean that a living body is so organized that everything has its teleological function; and this leads us to a proposed distinction between living and non-living bodies, namely, that living phenomena are essentially teleological. In the case of a machine we have the organization, but the teleology must be sought for in the living being that assembled the machine. Apparently, teleology is a characteristic of life; but yet everything is explicable on a physico-chemical basis; therefore we have in life the property of apparent teleology as a distinguishing characteristic. Only in this form can the proposed differentiation on the basis of "organization" be tenable. But, as we have seen, apparent teleology is one of the characteristics by which a reversal of the second law of thermodynamics can be recognized. It therefore follows that, in all probability, our distinction on the basis of the second law of thermodynamics is really the fundamental point of difference between living and non-living bodies.

Another suggested method of differentiation is in the capability of reproduction. But, when we come down to the ultimate living units, the cells, this reproduction consists merely of constriction and division; in which it is hardly to be differentiated from the breaking up into smaller drops of a drop of oil in water or a drop of mercury on a glass surface under slight shock. As we have seen, while under ordinary circumstances a shock is necessary to accomplish this division in these cases, yet, under the reversal of the second law of thermodynamics, this form of division is a normal phenomenon.

A further suggestion as to a method of differentiation is that life is always derived from other life, while inanimate matter may be derived from either living or non-living bodies. This distinction is a general one, simply stating a fact, but cannot serve as a definition or as a means of differentiation, because it would not show whether any individual case was one of living or lifeless substance. Should we try to apply the test, we should have to ask whether it could only have been derived from other living matter. What it could have been derived from we cannot experimentally find out; the actual causes might be discovered, and then we are reduced to the question whether life is to be found among those causes, and we are now no better off than at first. It is like trying in an unknown region to find the east by the directions in Schedrin's story: Face the north, and the east is on your right. Such directions obviously are useless where the north is as unknown as the east. The basis of fact behind this proposed distinction between living and lifeless bodies, however, we will examine more in detail later on.

The suggestion that organic bodies grow by absorbing particles, while growth, where it is found among inorganic bodies, is always by accretion of matter on the outside, turns out, when analysed, to be rather a distinction between solids and liquids than one between living and lifeless substance. The absorption of particles can be duplicated in the laboratory under certain circumstances by liquids enclosed in membranes, and a living cell consists of a membrane containing liquids.

Finally, we come to the dynamical distinctions. The most obvious of these is, to say that life is distinguished by movement. This is obviously an incorrect distinction, since all objects are in motion. But there is obviously something peculiar about living movement that seems to make it seem more mobile than other movement. It is thus, for instance, alleged, that living movement comes from internal causes, or else that living bodies work of themselves, while other objects need to be supplied with energy. Even that is not descriptive, for there are always "external" causes for all movements, and life does not create energy; if it uses up energy, it must obtain that energy from somewhere. Similarly with the distinction between static equilibrium of lifeless bodies and the socalled "dynamic" equilibrium of life, often more accurately defined as the metabolic process; such a dynamic equilibrium exists (as molar energy) in the case of almost all machines, and chemically in the case of any catalytic agent, which is also being constantly decomposed and recomposed.

But there are more accurate definitions of this mobility which is so peculiarly characteristic of life. We may notice, for instance, the theory advanced by the late Prof William James, the theory of the existence of a "reserve energy" in the case of biological, and especially in psychological, activities, which is absent in the case of lifeless activities. According to this, while the living organism can normally use a certain amount of its energy; yet in some mysterious way it can, under special circumstances, draw on an immense surplus fund of "reserve energy." This property being absent in physical bodies, we may draw a distinction on that basis between living and lifeless bodies, and this would seem to be an absolute distinction. Now, it has long been known that physical bodies contain an immense amount of energy which is unavailable for conversion into anything else; and the physical law that limits the amount of energy which it is possible for a physical body to utilize is precisely this second law of thermodynamics that has given us so much trouble. We must therefore come to the conclusion that, since life does not create energy, and this "reserve energy" is evidently real physical energy, that the peculiarity of life is its ability to draw on more energy than the second law of thermodynamics would allow; that is, its ability, in some circumstances at least, to reverse that second law. And again, we have seen that reversals of the second law are characterized by ability to use a fund of reserve energy that physical bodies cannot use. Let us say that the mechanical efficiency of a set of bodies is 85%; the reciprocal, or 118%, is that of the same set in the reverse universe. But as, under some circumstances, producing special results in the way of heat, etc., not quite 85% of the energy will be used, but, let us say, only 50%, then under those special cases in the reverse universe requiring more energy, the mechanical efficiency will be not 118%, but 200%, thus using over five times the amount of reserve energy normally used. This excess constitutes James's "reserve energy."

Another definition of the mobility of life is what is called "irritability," that is to say, the ability to make a large response to small stimuli. This, it is alleged, is possessed only by life, so that life may be defined by irritability. Against this Verwom objects that such inanimate substances as nitroglycerine also possess this property, that substance producing a powerful explosion under the influence of a slight shock. But in the case of nitroglycerine, we have an unstable equilibrium, and a slight shock simply lets loose the difference of level necessary to reduce to a stable equilibrium; while in the case of life, irritability is part of the so-called "dynamic equilibrium" and does not disturb that equilibrium. Irritability, as it is found in biological phenomena, is the ability to produce normally a large effect from a small stimulus without an irreparable leveling down of energy; in other words, the irritability that distinguishes life consists of the ability to build up higher differences, of energy-level from lower ones, in exactly the inverse order to that required by the second law of thermodynamics. In other words, irritability is identical with the "negative tendency" or, in other words, with the reversal of the second law of thermodynamics. Thus we are again reduced to our form of distinction between living and non-living bodies, namely, that between the negative and the positive tendency.

Verworn proposed the distinction on the basis of chemical constituency namely, that living bodies consist of complicated carbon compounds, such as albumen, protein, etc., which cannot be produced outside of life. But in what way would this definition distinguish a living body from, let us say, a corpse? Or, according to the definition by chemical composition, every wooden object is alive. It is obvious, therefore, that this distinction is untenable.

On the contrary, we have the extreme mechanistic view, represented by Dr. Jacques Loeb, that such a distinction cannot be drawn. The actual existence of a hard and fast distinction of this sort is, indeed difficult to prove, but there is certainly a difference in appearance, which must be based on something, however flimsy that something might be. Dr. Loeb calls a living body "a chemical machine," and states as the only base of differentiation "the power of automatic development, self-preservation, and reproduction." It is not quite clear whether or not all three properties are essential; and not all living bodies possess at all times all these three properties; while, on the contrary, these properties separately are possessed under certain circumstances by certain non-living bodies; so that, to say the least, this attempted distinction must be cleared up somewhat before it can be of any service at all.

Thus, of all the distinguishing characteristics that may be used to define life, we have left simply these three: apparent teleology, reserve energy, and irritability. The latter property (irritability) is, as we have seen, a condensed statement of the reverse of the second law of thermodynamics; while we have seen before that the other two properties, apparent teleology and reserve energy, are the outstanding characteristics by which a reversal of the second law of thermodynamics can be recognized. It follows, therefore, that the fundamental definition behind all these is: Life is a reversal of the second law of thermodynamics. Or, to put it in other terms, since we have seen that mechanical efficiency under the positive tendency is less than 100%, under the neutral tendency just 100%, and under the negative tendency more than 100%, we may define: Life consists of bodies with a mechanical efficiency of over 100%.

## VIII  THE EXTENSION OF THE SECOND LAW

We have, then, come to the conclusion that the second law of thermodynamics is not true as a general property of matter. It will, according to our theory, have to be omitted from the list of the physical laws. But what is there that we can put in place of it? We can say, in the first place, that every physical law is reversible, or rather, to be more accurate, that if any physical law is true, its reverse must also be true.

Furthermore, taking the conception of "mechanical efficiency," the second law of thermodynamics, if true, would set an upper limit to the amount of energy a body can use; namely, whatever difference in energy level there is. Now, if this second law is omitted from the list of physical laws, there is no such upper limit, there being more energy than that in the bodies, an inaccessible fund into which all energy tends to leak. But we can still use this limit, and express the amount of energy used by the body as a percentage of this limit. If we consider that the second law of thermodynamics is no longer a general physical law, this percentage may be 100% (neutral tendency), or less than 100% (positive tendency), or more than 100% (negative tendency). The mechanical efficiency of a body may thus fall into any of the three categories, but we have seen that 100% is a critical point, and when the mechanical efficiency changes from less than 100% to more than 100% or vice versa we have a change in the appearance of the action of the body. This critical point of mechanical efficiency constitutes the dividing line between living and non-living phenomena.

As we have seen, a universe following the positive tendency cannot have existed for an infinite time past; and the reverse of this rule must also be true, that a universe following the negative tendency cannot continue to exist for an infinite time in the future. Hence, if we suppose the universe to have existed from eternity past to eternity future, it follows that the average mechanical efficiency of the universe, taking all parts of space and time, must be exactly 100%. On the contrary, in our section of space and time, though we have found instances of the negative tendency (that is to say, life), yet the positive tendency visibly prevails, so that the mechanical efficiency at present of our part of the universe is considerably less than 100%. As the probabilities are that in our part of space for all time, or at the present moment for all space, the mechanical efficiency of the universe is about 100%, it follows that there must be other parts of space and time in which the mechanical efficiency is over 100%; and such parts of space and time supply us with examples of a reverse universe.

We have seen that the positive tendency is characterized by a constant running down of energy levels and a storage of energy into an inaccessible reserve store, which can in its turn be utilized and built up again into differences of energy-level only by the negative tendency. That is to say, the positive tendency stores up reserve energy, and the negative tendency once more utilizes it.

Since most of the substances within our observation follow the positive tendency, we may obtain characteristics of the two tendencies to some extent by observation, using our observations for the positive tendency, and reversing for the negative tendency. For instance, it has been observed on the earth that there is constant dissociation of atoms going on, especially in the case of substances of very great atomic weight (e.g., uranium and radium). It must be supposed that these substances must have been there in the first place in much larger amounts than at present, when the earth was in a hotter condition than at present; and accordingly we might expect, in very hot bodies such as the sun and stars, to find many substances with large atomic weight and few with small atomic weight, and in nebulas and newly formed stars to find substances almost entirely with large atomic weight, and almost no such substances as hydrogen, helium, etc., whose atomic weight is very small. The contrary, however, is true. In the sun, there is very little to be seen of substances of very large atomic weight; even such a substance as gold, which is more stable than uranium or radium and much more common on the earth, but with a large atomic weight, is conspicuously absent, while hydrogen and helium are present in large quantities (helium was first discovered in the solar spectrum as its name indicates). Furthermore, the nearer a star is to the nebular stage, the more conspicuously is this true; while in nebulae, temporary stars, etc., hydrogen, which has the very lightest atoms of any known substance, constitutes most of the substance of the star or nebula. It thus follows that in such hot bodies as stars and nebulae, there is an opposite process going on, which we may call the integration of atoms, the building up of larger atoms out of smaller ones.

All this is occurring under the positive tendency. If we suppose a section of the universe (either in space or in time) in which the negative tendency prevails, the reverse will be true. The integration of atoms will take place at lower temperatures, the dissociation at higher temperatures. It follows, that, if we consider a cycle in time of a body (or rather, of a large group of bodies), or of a spatial section of the universe, while we have two stages of mechanical efficiency, first building up reserve energy, then using up that same reserve energy for available energy, we have in a corresponding period a cycle of four stages in the evolution of atoms. In the first part of our "positive" epoch atoms are being built up, into more and more complicated forms; in the latter part of our "positive" epoch they are dissociated once more; during the beginning of the "negative" epoch the atoms are reintegrated, until sufficiently large differences of heat-level are built up to reverse the process, and the atoms become once more dissociated. Why this process should take place in just this way 1 cannot attempt to explain; but it may easily be that both dissociation and integration of atoms is constantly taking place, and the excess of one over the other would differ under different circumstances.

However, be that as it may, under the neutral tendency there would be no tendency whatever for the ultimate particles of matter to form into bodies or compound particles; so that we may expect that, even should the neutral tendency be found to exist, that there would be no "neutral" bodies, but that it would be entirely whatever the ultimate particles may be; e.g., that it would consist of separate electrons, if, as is at present believed, the electron is the ultimate material particle. Either the positive or the negative tendency starts out by building up more and more complicated atoms; but the neutral tendency does no such thing; accordingly we can take this as one of the characteristics of the neutral tendency. Another characteristic of the neutral tendency would be that, though it requires impenetrable matter, yet, since any friction resulting from the motion of bodies through it would tend to be counterbalanced by the equal negative element of building up motion, the result would be, an apparent lack of resistance, characteristic again only of the neutral tendency. And, inasmuch as it is now supposed that, though radiant energy is vibration transmitted by the ether, yet it is electrons scattered through the ether that are in vibration, and since the ether with its supposed electrons seems to be the only thing known that offers no resistances to the passage of moving objects, it follows that the ether, or the electrons it contains, is the example of neutral tendency to be found in our universe.

But we have said before that the probability of the neutral tendency is zero; how then does this come? As we have said before, a zero probability is merely an extreme improbability, but not necessarily an impossibility. For instance, if we have a finite segment of a line, and we select a point on the line at random, the probability that that point will be the middle point is precisely zero, since there are on the line an infinite number of points, of which only one is the middle point; so that the probability of the selected point being the middle point is 1 divided by infinity, that is, zero.

    C
A___ |___B

Or, to take another example, the probability that a point selected in space at random will be within the earth is: the earth's volume divided by the volume of space, which is zero since the latter quantity is infinite. Here, there are an infinite number of possibilities of the point being within the earth, and yet the probability is zero. Thus it is with the "neutral tendency." Its probability is zero, and yet there is a chance for an infinite amount of matter in the universe to come under it, provided that there is infinitely more matter that is either the positive or the negative tendency. In fact, we know from the theory of error that a mechanical efficiency of 100%, being exactly the average of the universe, is more probable than any other given mechanical efficiency, let us say 85%. And yet, in spite of that, its probability is zero. In fact, if we figure out the probability of the position of the universe at a given moment having come out exactly as it did, we also arrive at the conclusion that the theoretical probability of t he universe being as it is, is 0. And yet the universe exists, in spite of its zero probability.

Now let us consider the chemical structure of the positive and negative tendencies. For this purpose it will be necessary to distinguish between exothermic and endothermic compounds, that is, between compounds on a lower level of chemical energy than their constituents and compounds on a higher level of chemical energy than their constituents. We might expect that the former would be built up under the positive tendency, and the latter under the negative tendency, because the composition of the former out of their constituents involves the conversion of some chemical energy into heat, while the composition of an endothermic compound from its constituents involves the conversion of heat into a higher level of chemical energy. However, we must draw some distinctions here. The positive and the negative tendencies merely tend to build up respectively exothermic and endothermic compounds. Under special circumstances exceptions can be found.

It is true, indeed, in general, that the negative tendency tends to build up more and more endothermic substances; though many exothermic substances may result, either from more exothermic substances or where a substance, on account of its exothermic properties, has very little chemical activity. Also, the positive tendency does, as a whole, build up exothermic substances, though endothermic substances may be produced, usually as a result of a difference of energy-level higher than that of the endothermic compound. However, whether the negative tendency, for instance, builds up endothermic substances in negative or in positive objects, is a question which cannot be considered until we consider a little in relation to the reactions on one another of the positive and negative substances. We will accordingly proceed to investigate that.

## IX THE RELATION BETWEEN THE TENDENCIES

It would seem that the negative tendency depends for its possibility on certain special combinations of position taking place, which combinations would probably take place anyway by accident, but which would be much more likely to happen as a result of other similar combinations. In other words, if we take the negative tendency, we will find that any "negative" event must have been immediately preceded by an extremely improbable sort of combination, but is followed by a more probable combination. This can be seen if we take the simplest example of the negative tendency, the super-elastic collision, which, in order to happen at all, must be preceded by a rather unlikely sort of concentration of particles and energy at the exact point of collision. This does not in the least contradict our conclusion that the positive and negative tendencies are equally probable; for, on analysing the positive tendency, we find that it is followed by a similarly improbable condition. Thus, in the negative tendency, the cause is where the improbable stage comes in, but in the positive tendency, that same improbable stage comes in at the effect. However, the two varieties of "improbable stage" do not correspond exactly; for that of the negative tendency consists in a concentration of motion, while that of the positive tendency consists of a divergence of motion. Hence the effect of a positive event could hardly serve as the starting-point for a negative event. Outside of some accidental combination, a negative event must have a negative combination for at least part of the cause. The reverse of this rule is, that a positive event must give rise to positive effects, at least partially.

In other words, we have such things as a negative or a positive event giving rise to another event of its own kind; but with only positive causes, a negative result would hardly be expected to arise. If we identify the negative tendency with life, the statement reduces to this: All life comes from some living cause.

On the contrary, there is no such improbability in a purely negative cause giving rise to positive effects. In fact, as we have seen, a positive universe could not have existed for an infinite time past, nor a negative universe for an infinite time in the future; in either kind of universe, the change from negative cause to positive effect must take place; in fact, it is to be expected that it will be very co mm on for a negative cause to give rise to a positive effect.

We thus see that the transformation from positive to negative takes place in a very different way from the change from negative to positive. The latter can take place as a comparatively sudden transformation, a sudden cessation of all life activity; while non-living bodies cannot become alive except by accretion on other living bodies. The transformation from positive to negative can occur only as an extension of the negative tendency from some sort of center that is already negative; that is, by a living body growing.

It might be supposed that this difference between one kind of transformation and its inverse indicates an irreversible law; and we have already seen that, if we give up the second law of thermodynamics, we must replace it by the statement that all physical laws are reversible. Hence it would seem as though we had arrived at an inconsistency. But, if we examine into the question, we will see that one form of transformation is not the actual reverse of the other, but that each process is symmetrical in time, and is really the reverse of itself For the transformation, for instance, of a negative cause into a positive effect suddenly and completely, is a strictly reversible one, if we consider the fact that a negative cause corresponds in the reverse universe to a positive effect, and vice versa; so that, when in the real universe we have a negative cause and a positive effect, we will have the same in the reverse universe, so that the process remains unchanged when reversed. The same is true of the other process, by which a positive cause might give a negative effect.

However, in the latter case, there is a different element, whose reverse is not quite identical with itself, and therefore whose reverse can be used to supplement the proposition. That is to say, where such a transformation must, as we have seen, require some negative element to enter into the cause, the reverse of this requires some positive element to enter into the effect. That is, such a transformation not only cannot be spontaneous, but it also cannot be complete. If a positive substance be absorbed into a negative body, some positive matter must, at the end of the process, be rejected.

In other words, we come to the following conclusions: (1) Life cannot generate spontaneously, except by an accident that is so extremely unlikely that it would hardly happen once in a whole universe; (2) life extends to new matter by a process of growth, that is, by accretion round a living center; (3) where a living body absorbs inanimate matter, some inanimate matter must be rejected; (4) however, the transformation of living into lifeless matter may take place suddenly and completely, manifesting merely a sudden cessation of life-activities, a cessation which would be an irreparable one.

## X  EXOTHERMIC AND ENDOTHERMIC SUBSTANCES

To come back to the question that we began to consider, and which we left off in the middle; namely, that of the sort of chemical substances that would be built up under the two tendencies. We will have to distinguish between the case where we are dealing with a positive section of the universe and that where we are dealing with a negative section of the universe. To take the former case first, let us suppose that the positive is the prevailing tendency. This tendency would tend to build up exothermic substances, which the comparatively few cases of the negative tendency would form those same substances into endothermic substances for their own constituency. Some of these endothermic substances, it is true, will be rejected as positive or inanimate matter, but, on the whole, there will be a tendency for the more endothermic substances to go into the negative tendency, or into life, and for the more exothermic substances to be found in lifeless matter.

Since each process must chemically build up substances from their elements, which existed as free elements when the world was at a high heat, it might be expected that there might be a tendency towards complex compounds, so that substances tend, to a great extent, to combine with the tetravalent elements, which form the most complex compounds. The two most common tetravalent elements are carbon and silicon, the complex compounds of silicon being extremely exothermic, while the complex compounds of carbons are extremely endothermic. It might therefore be expected that inanimate matter would tend to build itself to a great extent into complex silicon compounds (silicates, such as earth, clay, many rocks, etc.), while, on the contrary, living matter might be expected to form as much as possible into complex carbon compounds, as endothermic as possible. Such is known to be the case; in fact, such carbon compounds are generally known as "organic compounds."

Furthermore, one substance that form s compounds of a high chemical energy, though itself having very low chemical energy is nitrogen. This element form s extremely endothermic compounds, which are in many cases explosive. At every ordinary chemical transformation involving nitrogen, some free nitrogen goes off into the air; but the reverse process, the fixation of nitrogen, that is to say, the formation of nitrogen compounds from nitrogen itself together with other necessary substances, is a process requiring an i mm ense amount of energy (by one process, a temperature of about 3000 degrees, by another process, a pressure of about 200 atmospheres). Since nitrogen form s such extremely endothermic compounds, we might expect that, where the general tendency is positive, life will tend to include not only as much carbon as possible, but also as much nitrogen as possible. It would therefore, in a section of the universe where the positive tendency prevails, seem to follow that life would tend, as far as possible, to be found in complex carbon-nitrogen compounds. The simplest of these compounds of carbon and nitrogen, itself an endothermic compound, is cyanogen, $(CN)_2$, and we might expect that the CN radical would be the foundation of life.

On the contrary, where a living body reacts with an inanimate body in any way, it is also likely to build up such complex carbon-nitrogen compounds not only as the living product, but also as the lifeless product which we have seen must be formed. Hence these products must be formed to some extent not merely in living matter, but also in inanimate matter. For instance, this very process of the fixation of nitrogen, that we have already referred to, we might expect to be found accomplished by living bodies which can absorb nitrogen and react with it, leaving nitrogen compounds as rejected matter, besides forming themselves into nitrogen compounds. We do, in fact, find such a process operating among what are called the nitrogen-fixing, or nitrifying, bacteria, which absorb nitrogen and reject non-living nitrogen compounds in a manner that could hardly be explained as anything but reversing the second law of thermodynamics.

Thus is the result where the prevailing tendency is positive, and where the negative tendency is the exception. To trace this result further, we much remember that life, the negative tendency, grows by accretion on a living center which is necessary. Living bodies absorb inanimate matter, extending life more and more, absorbing to some extent exothermic substances, rejecting to some extent endothermic substances, until this living activity begins to take in the majority of the section of the universe. Meanwhile the living, the negative, activities will have absorbed most, if not all, of the exothermic substances, while the positive tendency will be kept up by the constant rejection of mostly endothermic substances as lifeless matter. Thus will the extremely complicated carbon nitrogen compounds tend, in a section of the universe where the prevailing tendency is negative, to be found more and more as positive, as lifeless bodies. Furthermore, since such a section of the universe is the exact reverse of a positive section of the universe, such positive bodies will tend to be formed as exactly such complex organisms as are, in our section of the universe, found in living bodies. We will have a complex, life-like organism, but with none of the life activities (with some exceptions, as we shall see). We may call such organisms pseudo-living organisms. In our "reverse universe" these pseudo-living organisms will take the exact shapes of the living organisms in our real universe.

Such extremely endothermic compounds are unstable under the positive tendency, but require the negative tendency to stabilize them. Under the positive tendency, these compounds will tend to decompose into exothermic tendencies very quickly. But the tendency of negative activities to extend from a negative center will be very active when most of the universe is negative, and hence such exothermic substances will be likely to be quickly absorbed by the prevailing negative tendency; while, on the contrary, the prevailing negative tendency will tend quickly to build up as rejected positive matter these same endothermic compounds into the positive, pseudo-living organisms. Thus these pseudo-living organisms differ from corpses in that there is a constant cycle of chemical reaction with the surrounding world, a constant building up and decomposition of substance. Since these organisms are the exact reverse of living organisms as we know them, it follows that, in a section of the universe where the prevailing tendency is positive, any living bodies much exist in the form of chemical machines that constantly absorb inanimate matter, build up into living matter, and as constantly make partial decompositions of their own substance into more exothermic substances which are rejected as inanimate matter. That is, both living substances in our section of the universe and the pseudo living organisms in the negative sections of the universe have in co mm on the property of metabolism. All these conclusions hold except at a heat so great that the formation of compounds is impossible (e.g., on the sun). Metabolism is thus not a property of life, but of the minority tendency. The same is true of the chemical composition of organisms. In a positive section of the universe the organisms are living; in a negative section of the universe they are essentially lifeless.

Where the heat is too great to permit of the formation of chemical compounds, such chemical machines cannot exist; but the minority tendency, whether positive or negative, would probably exist, the chances of its nonexistence being extremely small. Under any conditions the chances are overwhelmingly in favor of there being a mixture of the two tendencies. Yet, though both tendencies are present, there will be a majority and a minority tendency. But what such minority tendency may be like, it is difficult to imagine. For instance, it would be difficult to imagine what sort of phenomenon life would be on the sun. It would certainly have to be different from any life that we know of, though with the common properties of irritability, apparent teleology, and reserve energy.

## XI THEORIES OF THE ORIGIN OF LIFE

According to our hypothesis, life always has existed and always will exist under all conditions in some form, though that form may be quite different from any form of life that comes within our experience. If we trace back the ancestry of present-day life, we will always be able to trace it back to some life, though it may be in such a form that it might be extremely difficult to recognize it as life.

Thus, there never was a time when life started on the earth; it merely developed into its present complex form from some simpler form that existed on earth when the earth was in a molten or even in a vaporous condition; still further back, it can be traced to some extremely simple form of life that existed as far back as the nebula out of which the solar system originated; we shall later attempt to trace it back beyond the nebula.

Our theory of the origin of life is thus that there was no origin, but only a constant development and change in form. This belongs to the class of theories known as the Biogenetic theories, as contrasted to the

Abiogenetic theories, which assume that at some previous time life did not exist, and that under certain special circumstances that existed when the earth was in a heated condition the necessary elements came together somehow and assembled themselves into a living body from which all other living bodies are descended. The nature of this automatic assemblage of constituents remains, of course, rather mystical; not to speak of the fact that the assumption of spontaneous generation is rather contrary to observed facts.

Such abiogenetic theories have very frequently been advanced, especially since every now and then there is a recrudescence of the belief that spontaneous generation of life is possible under present circumstances, that life could be produced in the laboratory, in spite of all observed facts to the contrary. Haeckel represents this abiogenetic theory in its most general form: that, when water first liquified on the earth, its reaction on various substances then present produces proteid, which was the original life from which all life has descended.

How this proteid was formed remains a mystery. We have, however, a more detailed explanation in Pfliiger's theory, which is to the general effect that the original combination was that between carbon and nitrogen, forming cyanogen, which in its turn united with other substance, especially the hydrogen and oxygen of water, to form more and more complex cyanogen or other similar carbon-nitrogen compounds, such combination forming as a final result proteid, the chemical basis of which would thus be the cyanogen radical, CN. Here we have a very likely explanation. In the first place, even many simple cyanogen compounds have many chemical reactions very similar to those of proteid; in the second place, whereas proteid has, in itself, been found to constitute a very great poison, it is also known that cyanogen is one of the most powerful poisons known, and that its compounds are, in general, extremely poisonous, though in certain combinations found in living bodies (e.g., almonds) such compounds seem to be quite harmless.

One difficulty with this is that this idea of spontaneous generation is somewhat contrary to observed facts; there is no instance observed of such a thing as spontaneous generation of life. But the main difficulty is to explain the formation of cyanogen and especially of its compounds from their elements. It is perfectly true, as alleged, that carbon and nitrogen being together at a high temperature will form a little cyanogen, but, with the large amount of oxygen present, this cyanogen could not last long, since cyanogen is a very endothermic substance, and under the second law of thermodynamics, it will reduce to the combination that has less chemical energy, losing the difference in the form of heat; resulting in carbon dioxide and nitrogen as a product. Accordingly we must suppose some peculiar sort of carbon and nitrogen that will not only unite into cyanogen, but which will form such a peculiar form of cyanogen that, instead of oxidizing on contact with oxygen (as ordinary cyanogen would), it not only holds itself aloof from oxidation but even form s more complex and more endothermic compounds. We must suppose some form of carbon and nitrogen which would reverse the ordinary chemical reactions under those circumstances; or, since those reactions are based ultimately on the second law of thermodynamics, we must suppose that there was at that period of the earth's history some carbon and nitrogen that possessed the ability to reverse the second law of thermodynamics. If we suppose that, our theory of life can easily harmonize with the Pfliiger idea as to the origin of organic life from cyanogen compounds. In fact, as we have seen, our theory of life is such that we would theoretically suppose that living organisms would have a chemical construction based on the cyanogen radical, thus falling in exactly with Pfliiger's idea that life on the earth originated in cyanogen and its compounds.

So much for the abiogenetic theories. Turning now to the biogenetic theories, we can hardly find them much more satisfactory. We have, for instance, Preyer's theory that the earth itself, in the heated state, was itself an immense living organism, from which all living organisms existing at present are descended; all inorganic matter on the earth being merely the rejected excretions of the former living earth, while the living substance came more and more to resemble protoplasm.

Absurd as this theory may sound, there is nothing impossible about it. However, the astronomy of the proposition is rather poor. There is no reason to believe that the earth, in a heated state, was in any different condition from all other known bodies which we find in a similarly heated state (e.g., the major planets and the sun); and these heavenly bodies are hardly in a condition which could by any stretch of imagination be called living. However, if by living is meant that there is a potentiality of the generation of life, of course the earth in a heated condition must have come under that heading. But since the heated planets have no particular resemblance to life, what is more likely is, that on the earth in a heated state, there was life, and there was such life even in the nebula, almost as different from the life that we know at present as the kind of earth-organism that Preyer supposes, but having some properties in co mm on with present life. This is precisely what our theory of life would lead to.

We come now to the biogenetic theory most commonly advanced, one which numbers among its supporters Helmholtz and Sir William Thompson. This is the so called theory of cosmozoa[1], otherwise known as the theory of seed-bearing meteors. This theory is to the effect that two planets, at least one of which had developed life on it, came into collision, so that each of the planets was broken into small pieces which were scattered all over in different directions. Some of the pieces of the

1      cosmozoa hypothetical seeds or spores that led to the beginning of life, said to have reached Earth through space.

life-bearing planet, in the form of meteorites, passed another, always bearing within them the seeds of life. These meteorites finally came into the solar system and entered the earth's atmosphere, then striking the earth and planting these seeds of life on the earth, which afterwards developed into the various form s of life that now exist.

This theory sounds very plausible, but again it is based on very poor astronomy. The co mm on meteors, or shooting-stars, which are the actual bodies that seem to have passed from one system to another in this manner, never actually reach the earth's surface, but are completely burned up before they have penetrated very far into the atmosphere. The larger bodies that actually reach the earth's surface are the so-called meteorites which are as much parts of the solar system as the planets, and even move round the sun in the same direction as the planets; which are, in fact, simply stray asteroids. Tracing life to such bodies is not tracing it to any other stellar system, but merely to the solar system, and makes it more impossible than ever to trace back where these supposed seeds of life came from, or how they got into the meteorite. There is, of course, nothing to prove that such a meteorite as this hypothesis assumes could not be formed, or that it could not thus transplant life from one planet to another. But, since such meteorites as are large enough to reach the earth's surface and at the same time are not members of the solar system do not seem to be a regular occurrence, there being no known instance of such a body, it would seem that such an occurrence is a very rare one, if indeed it can happen at all that any life in its present form could survive such a collision or such a long trip through space. Should two planets happen to collide as the hypothesis assumes, and should any life on those planets survive the collision, the chances are almost nil that, in the case when it happens, any of the pieces of wreckage would strike another planet at all, much less that it would strike one at the very period when that planet was ready to receive the very form of life carried by the meteorite. Thus not only is the hypothesis improbable per se, it is also contrary to any observed facts, since the actual bodies that could be supposed to come from other stellar systems are, as far as observation goes, so small that they are burned by friction with the atmosphere long before they can reach the surface of the earth.

It might be interesting to note that, in the time of Helmholtz and Thompson, the distinction between the meteors or shooting-stars that come from other stellar systems, and the meteorites or aerolites which form part of the solar system, was a distinction which had not yet been clearly drawn.

We thus come to the conclusion that the theory of Cosmozoa is entirely unacceptable in view of present facts, while Preyer's theory of the earth-organism can only be accepted in the extremely modified form that the earth, in its molten and vaporous states, contained life (instead of having been alive, as Preyer himself would have it). Thus we come to the conclusion that life is as eternal as the inanimate, and is to be found as universally, under as varying conditions, as inanimate phenomena. On the other hand, we can also accept the Pfliiger idea that life as it exists on this earth originated from the formation of cyanogen and its compounds.

y of the origin of life is really Biogenetic, in that it supposes that all life originated from life for an eternity past; but, on the contrary, inasmuch as the past life from which the present life is derived was in an almost unrecognizably different form from that in which life at present appears, it supplies a basis for the Abiogenetic theories of the origin of life from non-living organisms which are, according to our theory, inorganic life.

## XII  THE ASTRONOMICAL UNIVERSE

The consideration of the question of the origin of life and the various theories formed on that question has led us into astronomical considerations, so that it may be worth while to examine the astronomical aspects of our theory of the reversibility of the universe. And we may well, after dealing both with objects of ordinary size and with very small and even ultimate particles, turn to the consideration of objects of a different, a larger scale of magnitude: the heavenly bodies. We shall therefore consider our theory in connection with such objects. Astronomy deals not only with individual planets in our solar system as a whole, but also with the almost inconceivably vast extents of space which stand between the various stars and their special systems, and, finally, with the theory of that general group of all stars which is known to astronomers as the stellar universe, or simply as the universe. Accordingly one of the first things we should investigate should be the astronomical theories of the universe, especially since our theory of reversibility is essentially a theory of the universe.

When we come to examine the astronomical theories of the universe, we find that they divide themselves into two groups. Just as the biological theories of the nature of life are, generally speaking, to be divided into the mechanistic and the vitalistic, so the astronomical theories of the nature of the universe may be divided into the theories of the finite universe and the theories of the infinite universe. And, as our theory effects a compromise between the two kinds of theories of life, we may try to see whether our theory cannot also reconcile the two kinds of theories of the universe. Let us, therefore, examine in more detail each of the two kind"s of astronomical theories of the universe and the various arguments that can be adduced in support of both kinds of theories.

Let us take the theories of an infinite universe. The general idea of these theories is, that space is infinite, and there is no special reason why matter should be confined to one portion—and, at that, only an infinitesimal portion compared to the infinity of space. Thus we get the picture of an infinite geometrical space filled with stars, here to a somewhat greater density, there somewhat less densely, but, on the whole, with a certain average density. This reasoning on the basis of the theory of probability is a perfectly good one, and it is, furthermore, not the only argument in favor of an infinite universe. There are arguments that are based not on theory but on actual observation.

The most important of these is the gravitational consideration. If the universe is infinite, and matter approximately uniformly distributed throughout the universe, then, on the average, the gravitational pulls on a given stellar system should, on the whole, completely balance each other, so that gravitation would not tend to pull any stellar system in any particular direction, and the proper motion of any star should be, in accordance with the law of inertia, a uniform motion in a straight line. But, on the contrary, if there were in the universe a center of density, and especially if there were a finite universe, then all stellar systems would tend to be pulled on towards that center of density and, in general, revolve round that center. The facts indicate that the proper motion of stars is actually uniform motion in a straight line, and that there is no center about which all stars move; so that this argument would point most distinctly to an infinite universe, with matter distributed throughout space approximately uniformly, one part of space being in this respect no different from another.

But there is one outstanding objection to this theory that the stellar universe is infinite. There may be supposed to be no reason why the average brightness of stars should be any different in one part of space from what it is in any other part; multiplying this average brightness by the average number of stars per unit volume (the average star density that we suppose for the infinite universe, we will get the average amount of light issuing from a unit volume anywhere in space; let us call this product L. Now, as the apparent brightness of any source of light is inversely proportional to the square of the distance between that source and the observer, then, if we call that distance d, the average apparent brightness of a unit of volume at distance d from the observer could be represented as $L/J^\wedge$. If we divide space into an infinite number of concentric spherical shells, with the observer at the center, each with equal thickness, let us say the unit distance divided by An, then, especially when the sphere is very large, the volume of each shell is approximately $d^\wedge$. Multiplying the average apparent brightness of a unit volume at distance d by the volume of the shell of distance d, we find that the volume of each such shell is a constant, 1. Since the stellar universe consists of an infinite number of such shells, each of which has the same apparent brightness, it follows that the brightness of the sky, or indeed of the smallest part of it, must be altogether infinite. The consequence of the theory of an infinite universe is obviously contradicted by facts.

On account of this objection to the universe being infinite, there arose the theories of the finite universe, which seem to depend mainly on the observed distribution of light in the sky (outside of the light from the sun, moon, and other members of the solar system). These theories of the finite universe started with the great observer. Sir William Herschel (one of the three originators of the Nebular Hypothesis), whose theory is that of the so-called "drum universe." According to Herschel's theory, the universe is in the shape of a very flat circular drum, or, in other words, a thin, wide circular slab, with possibly another secondary slab at a plane inclined a few degrees to the first, the two slabs being concentric, and the center being—the sun! It seems that even Herschel had the idea that our solar system is the center of all things, which is somewhat a survival of the ancient doctrine that the earth is the center of the universe. In fact, we may say that Herschel's theory of the universe is a modernized version of the ancient primum mobile[2] containing the stars and having the earth for a center. However, the drum (or double-drum) shape of the universe is intended to explain the distribution of light; for, in a plane of the drum, we should have to look through such an immensely greater amount of stars than in a direction with any considerable inclination to the plane, so that we should have the appearance of a white streak running all around the sky, which we actually have under the name of the Milky Way; the double-drum shape would require a bifurcation of this white streak at two opposite parts; which again is strictly in accord with observed facts. Herschel was perfectly willing to believe that there are other similar drum-shaped universes, two of which, according to him, are visible to us, and known as the Magellanic Clouds. These "clouds" were first discovered by the famous explorer Magellan, and are circular patches in the sky of the southern hemisphere which look like detached portions of the Milky Way, though at a considerable distance from the Milky Way.

The modem theories of the finite universe, though not accepting Herschel's explanation as to the Magellanic Clouds (rather tending to suppose that those objects are within our own stellar universe), are very similar to Herschel's drum theory in general outline, and all have the same characteristic of being attempted explanations of the distribution of light actually found in the sky. The tendency, however, is not to suppose that the solar system is at the center of the universe, but rather to suppose that the solar system is considerably south of the center, being almost on the southern side of the drum, and much nearer the southern part of the drum edge than it is to the northern. There is, further, a tendency to suppose that this stellar universe is the result of a collision of two semi

2    In classical, medieval, and Renaissance astronomy, the Primum Mobile (Latin: "first movable") was the outermost moving sphere in the geocentric model of the universe

universes, which is what we have seen would be the result of pushing the second law of thermodynamics to its logical conclusion, it being an observed fact that the stars seem to move in two general currents. However, just as the theory of the infinite universe cannot be supported on the grounds of the distribution of light, so similarly the theories of the finite universe cannot be supported on the grounds of the consideration of gravitational attraction.

We thus find that considerations of gravitational attraction lead us to suppose an infinite universe with stars approximately uniformly distributed throughout space; similarly with considerations of probability, which lead us to the same conclusion. But, on the contrary, the observed distribution of light in the sky leads us to the directly opposite conclusion, that our stellar universe is finite, though there may be stray stars outside that universe that occasionally come in, and though similarly some stars may occasionally stray out of the limits of the universe. There may be other such finite universes, in which case we may conceive of things in such a series as the following:

Electrons are the particles that make up atoms;
Atoms are the particles that make up molecules;
Molecules are the particles that make up masses;
Masses are the particles that make up planets, etc;
Planets, etc., are the particles that make up stellar systems;
Stellar systems are the particles that make up universes;
Universes are the particles that make up existence.

All of which sounds perfectly reasonable; but the gravitational consideration spoils this simple series; and it is a consideration that cannot easily be disposed of It would seem, then, as if there was gravitationally an infinite universe, while in relation to light the shape of the universe is something like Herschel's drum. In other words, stars are uniformly distributed throughout the whole of infinite space, so that the gravitational phenomena will be like those of an infinite universe; while somehow or other, beyond Herschel's drum, stars do not give out light. This phenomenon cannot be explained by a partial opaqueness of ether; for then the apparent shape of the universe would be spherical, with ourselves at the center, instead of double-drum-shaped, with ourselves on the southern side. Hence there must be some other explanation, especially since this same question of probability indicates that ether is likely to be uniformly distributed through infinite space.

Some other explanation, then, must be found. Beyond the boundaries of Herschel's drum, for some unknown reason or other, stars fail to give out light. Either they are all cold or they are hot but not bright. And furthermore, stars must be constantly entering and leaving the limits of this Herschel drum. We may easily suppose that a star, after having passed all the way across this part of space, has cooled down so much as to give no light; but on entering, they are much hotter than later on, because stars constantly lose heat to the surrounding ether; hence, if these stars were cold before entering the Herschel drum, something must have happened to them near the boundary to heat them up suddenly. If there is, around the boundary of the drum, any material which would heat up a star by collision, friction, or contact, then it would follow that cold stars leaving the drum would be similarly affected; which is hardly in accordance with the theory, as deduced from observation. Hence we conclude that the stars which enter the Herschel drum are, to a great extent at least, hot, but give out no radiant energy (light). Thus, outside the limits of the Herschel drum, as far as we can judge, stars exist, and many of them are even hotter than the stars within our observation, and it would seem that the ether is there to receive radiant energy from them, but no radiant energy is forthcoming.

The result, then, is, that we do indeed have an infinite stellar universe, but that Herschel's drum has the peculiarity that, within it, stellar heat is converted into radiant energy, while no such conversion takes place outside the Herschel drum. There may, furthermore, be other Herschel drums in other parts of space having similar peculiarities. In order to understand the special peculiarity of these Herschel drums, let us examine why stellar heat is converted into radiant energy at all.

In the first place, the ether of interstellar space is at a very low temperature, while, in general, a star is at an extremely high temperature, many stars being much hotter than our sun. According to the second law of thermodynamics, the energy should tend to run down towards a common level; that is, the star's heat energy would radiate into the surrounding space and appear in the form of ether-vibrations, that is, in the form of radiant energy, under which heading is included light. If, then, outside Herschel's drums, there are many hot stars, hot enough to give out light of all vibration-periods (white hot), but which do not issue any radiant energy, it follows that somehow the second law of thermodynamics applies only within the Herschel drums but is somehow suspended or even reversed outside them. In other words, the actual stellar universe, as manifested by gravitational phenomena, is infinite, and stars are approximately uniformly distributed throughout infinite space; but we can only see the stars in that section of space where the second law of thermodynamics prevails, and therefore the section of the stellar universe that is visible is, after all, only finite.

We thus come to the conclusion that the boundary of the Herschel drum is really the limiting surface between positive and negative sections of the universe. And now we come to the question whether, starting with our theory of the positive or the negative tendency prevailing in different parts of space and time according to the theory of probability, we can draw any more detailed conclusions in respect to the exact appearance of the stellar universe.

In the first place, we have come to the conclusion that, taking any given moment of time, the positive and the negative parts of the universe should be approximately equal, as a matter of probability; in fact that, if we take the whole of space and time, the positive and negative sections bear towards one another a ratio of exactly 1. Since we are dealing with only the present time (or times near the present) in dealing with the present appearance of the universe, we may confine ourselves to the statement that, in a given portion of time, there should be approximately equal positive and negative sections of space; and, if matter is approximately uniformly distributed throughout space, that the volumes of the two kinds of sections should be approximately equal. The next question is, in what way the negative section of space can be distinguished from the positive section.

Our previous consideration on the production of radiant energy from the stars indicates that such production of radiant energy is only possible where the second law of thermodynamics is followed, that is, in a positive section of the universe. In a negative section of the universe the reverse process must take place; namely, space is full of radiant energy, presumably produced in the positive section of space, and the stars use this radiant energy to build up a higher level of heat. All radiant energy in that section of space would tend to be absorbed by the stars, which would thus constitute perfectly black bodies; and very little radiant energy would be produced in that section of space, but would mostly come from beyond the boundary surface. What little radiant energy would be produced in the negative section of space would be pseudo-teleologically directed only towards stars which have enough activity to absorb it, and no radiant energy, or almost none, would actually leave the negative section of space. The peculiarity of the boundary surface between the positive and negative sections of space, then, is, that practically all light that crosses it, crosses it in one direction, namely, from the positive side to the negative side. If we were on the positive side, as seems to be the case, then we could not see beyond such surface, though we might easily have gravitational or other evidence of bodies existing beyond that surface.

Furthermore, just as, in the positive section of space, light is given out uniformly in all directions, so, in the negative section, light must be absorbed by a star equally from all directions. Thus, to any star in the negative section, light must come in about the same amount from all directions; and, since most of this light comes from the positive sections, it follows that the negative sections must be completely surrounded by positive sections and must therefore be finite in all directions. By reversing this (since we have seen that all physical laws are reversible), it follows that any positive section must also be finite in all directions, and be completely surrounded by negative sections. We thus find the universe to be made up of a number of what we may call bricks, alternately positive and negative, all of approximately the same volume; a sort of three-dimensional checkerboard, the positive spaces counting as white (giving out light), and the negative spaces as black (absorbing light).

Thus what we see is simply the white space that we are in. The surrounding black spaces are invisible, and in addition, absorb the light from the white spaces beyond, so that even those cannot be seen, and, if we judge from the distribution of light in the sky, we get an idea merely of the size and shape of our special white space.

Let us try, now, to get a theoretical idea as to approximately what should be the shape of these white and black spaces, so that it can be compared with observation. For developing the theory in this direction, we must remember that the proportion of positive matter in any part of space should, according to probability, be about 50%. But this same theory of probability will tell us that it is extremely improbable in any given part of space that this proportion should be exactly 50%, but that there should be a discrepancy between the percentage of positive and that of negative phenomena, this discrepancy becoming increasingly improbable the greater the discrepancy is. Accordingly we may suppose that there are surfaces where the proportion of positive events is 50% (our boundaries), and other similar surfaces where there are other special proportions, while, in the middle of the positive "bricks," there will be a maximum percentage point, and in the middle of the negative "bricks" there will be a minimum percentage point. Around these maximum and minimum points our white and black spaces will be built, the fundamental variation of the percentage away from these points being presumably based on three principal directions or dimensions, of which the variation in other directions will be compounded.

Proceeding from, let us say, one of the maximum points (center of a positive section of the universe) in any direction, the discrepancy from the normal of 50% should become first positive, then negative, in a sort of vibrationary form. This vibration should be irregular, according to the theory of error, though with a certain average; but in the three principal directions, approximately perpendicular to each other, we should expect to find them more uniformly periodic.

If these "vibrations" were regular and perfectly periodic in these three directions, the boundary surfaces would be planes midway between the maximum and minimum points, and the sections of the universe would take the shape of rectangular parallelopipeds[3]. With such shape, the sections of the universe would indeed be "bricks." But such regular uniform

3      In geometry, a parallelepiped is a three-dimensional figure formed by six parallelograms. By analogy, it relates to a parallelogram just as a cube relates to a square.

vibrations are hardly to be expected. The theory of error would lead us to expect irregularities from even that; but the volume of the sections should remain unaltered. Furthermore, a positive section must touch another positive section along an edge, or else at that edge two negative sections will form a continuous section, and we are thus liable to get a continuous line of negative space to perhaps an infinite extent, which is contrary to anything that we should expect. Hence we must expect that, in the irregularities, both the edges and the volume would be but slightly changed.

The faces of the parallelepiped, however, may, even under these conditions, be considerably changed. We may, for instance, expect that the vibrations of the percentage, instead of being the simple-harmonic vibrations which would produce plane boundary surfaces midway between the maximum and minimum points, may be compounded with its "harmonics," that is, may be compounded with vibrations of multiple frequency, of which the double frequency is the most important. The double frequency would be likely to make a whole face of the parallelopiped either cave in or bulge out, the higher frequencies will simply introduce further irregularities. Since there is to be little alteration of volume of the sections, two of the opposite pairs of surfaces must be changed in one direction, and the third in the other. The longer dimensions of the parallelopiped are those in which more irregularity is likely to show itself, so that the biggest alteration would show itself on one of the two smaller pairs of opposite faces. The other two pairs of faces will then have to be altered in the opposite way to make up for this; presumably the largest and the smallest, the medium pairs of faces showing the greatest irregularity. The irregularity may thus be of two varieties: either the medium pair of faces is caved in, and the largest and smallest bulged out somewhat less; or the largest and smallest pairs of faces are caved in slightly, and the medium pair of faces extremely bulged out.

Taking each of those two shapes (and they are liable to alternate to some extent, some sections of the universe being of one kind of shape, and some of the other), we can suppose of each one that it represented a positive section of the universe, and attempt to predict the distribution of light in the sky as seen from somewhere near the maximum point. If the parallelepipeds are comparatively flat (as they are likely to be, the three dimensions of these figures probably being widely different), it follows that in the sky, the plane parallel to the largest pair of faces would seem to be filled with a thick white strip. According to which of the forms of irregularities we suppose, the shape of the strip will vary. If the largest and smallest faces are bulged out, this white strip would be much less conspicuous, there being in other directions a good distribution of stars visible, but the strip would still be visible, and the hollow in one pair of faces would mean that, in one place on the strip, as well as in the opposite part, there would be a widening (due to the medium pair of faces being nearer than the smallest, and consequently, appearing wider) with a dark space in the middle of this widening. Midway between these dark spaces the strip becomes narrow, due to the fact that there the surface bounding the section of the universe recedes to a great distance. If the other shape of the positive section were adopted, we should have something similar, except that the strip would tend more to be of uniform width, and, if anything, the "coal-sacks" would be in the narrow part of the strip. We may represent the two forms of the strip somewhat as follows:

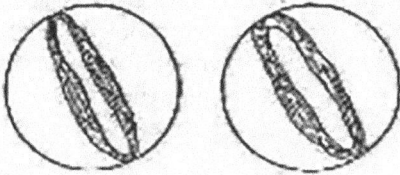

These "coal-sacks" would tend to be oval in shape, instead of pointed at the ends, as Herschel's double drum would lead us to suppose. If we are on the southern side of the positive section, then on the southern side more irregularities would be seen, such as striations of the strip, occasionally small "coal-sacks" in other parts than where expected, while some of the irregular wavy variations on the largest face of the "brick" on the south side would result in our seeing, near this strip, apparently detached sections, presumably approximately circular. As a matter of fact, the so-called Galaxy or Milky Way has the shape indicated in the first of the two above diagrams, with exactly such irregularities as we have predicted. The shape of the coal-sacks is indeed approximately oval, and not pointed, as Herschel's theory would lead us to expect. Furthermore, such circular detached sections of the Milky Way actually do appear in the southern hemisphere, and have been phenomena which have always been difficult to explain; they are called the Magellanic Clouds, and we can see that, according to our theory, they are exactly what they look like: detached sections of the Milky Way. And, if they result from what we suppose, namely, the largest of the three southern faces of the "brick" becoming wavy and extending suddenly a great distance out, it follows that the neighboring regions, which are the opposite phase of the same waves, should be so near us that there should theoretically, around the Magellanic Clouds, be very few stars visible. This is indeed the case; the Magellanic Clouds are found in a region of the sky that is almost completely devoid of stars.

We thus find that not only does our theory of a reversible universe actually reconcile the theories of the infinite universe with the theories of the finite universe, but it actually enables us to predict the distribution of light in the sky much more accurately than any theory has yet been able to do. We thus see that the universe is infinite, but divided into alternately positive and negative spaces of approximately equal volume, and that the apparent stellar universe is merely the positive section in which we are. The Galaxy consists merely of the distant sides of the irregular "brick" that constitutes this positive section.

To get an approximate idea of the size of this "brick," the temporary star. Nova Persei, which appeared in 1902, was in the Milky Way, and was probably as distant. Its distance has been estimated at about 3400 light-years, so that this gives us the length of the "brick" as about 7000 light-years. The Milky Way near the coal-sacks being about twice as wide as here, the width of the "brick" would be about 4000 light-years. And, the greatest width of the Milky Way being about 15 degrees, that gives the thickness of the brick at about 1000 light-years. In reducing to ordinary measurement, we may notice that a light-year is about 5.8 trillion miles.

## XIII  THE NEBUEAR HYPOTHESIS[4]

So far, we have considered only a single cross-section in time of the universe in our astronomical considerations. That is, we have only considered the appearance of the universe at a given moment of time, and thus come to the conclusion that, at a given moment, the universe is built of positive and negative sections with a tolerably well defined shape, as a consequence of our theory of the reversibility of the universe. But we have not yet considered the changes in the universe or in its constituent parts, the stars, that are brought about by time; what the universe was in the past, what it will be in the future; in short, the course of events that generates in individual stellar systems, or in the universe, its past, present, and future conditions. This branch of astronomy is known as Cosmogony.

Let us, then, examine the recent theories on the matter of Cosmogony. The first theory which had a scientific basis was the so-called Nebular Hypothesis. This hypothesis is interesting partly from the fact that it originated in the minds of three men independently at about the same time, these three men having arrived at it from three different points of view, and being in three different countries, while each was among the most prominent men in his own specialty. One of those three men was Immanuel Kant, the famous philosopher, who originated this hypothesis as an incidental speculative conclusion from his own philosophy; another was the well-known mathematician Laplace, who arrived at the hypothesis

4      It suggests the Solar System is formed from gas and dust orbiting the Sun which clumped up together to form the planets. The theory was developed by Immanuel Kant and published in his Universal Natural History and Theory of the Heavens (1755) and then modified in 1796 by Pierre Laplace.

from considerations of his studies of celestial mechanics; while the third originator of this hypothesis was Sir William Herschel, who is well known as an astronomical observer, and who arrived at the Nebular Hypothesis as an explanation of many phenomena he observed among the stars.

According to this theory, which was quite generally accepted almost throughout the nineteenth century, the universe was once what the originators of the hypothesis have been pleased to call (probably after the Greek mythology) a Chaos; which seems to mean undifferentiated matter uniformly distributed throughout infinite space. Centers of attraction were formed where, in any spot, the matter was slightly denser than in its vicinity, and surrounding matter was drawn in to these centers. Thus were formed whirlpools, which set up a great rotation at each center, increasing with the increasing condensation at the centers. It was further assumed that the original Chaos was at an intense heat, so that, at the centers of attraction, great, hot, rotating bodies were formed. The rotation became faster and faster as the matter was drawn in toward the center, the centrifugal force finally becoming so great that the rings of matter were thrown off. In each ring a center of attraction was formed, and the process was repeated. From the primary whirlpools there thus came the stars; from the centers of attraction resulting from the rings, came the planets; and these planets themselves threw off rings which finally became satellites. Since larger bodies cool off slower than smaller bodies, the stars and the largest planets remained hot, the smaller planets and the satellites cooled off to a solid condition.

This theory was accepted generally throughout the nineteenth century, with occasionally some minor modifications. For instance, the assumption that the original Chaos or universal nebula was in a state of intense heat had since been dropped, because the energy of matter coming in from a distance under the influence of gravitation would be sufficient to explain that heat would arise in immense amounts. When potential gravitational energy at a high level is reduced to a smaller amount at a proportionately lower level, the difference is converted to heat without loss under the second law of thermodynamics.

This theory was corroborated by the supposed fact that the planets and stars in the various stages were visible to astronomers; such as hot planets (the major planets, such as Jupiter and Saturn), and even the ring around Saturn, stars in the various stages from extremely heated stars to almost dark stars, Algol's companion-star being an example of a completely dark star; while the very first stage of star-formation would be indicated in the many nebulas that are visible in all parts of the sky. One stage, of course, that was not exemplified in observation, was the Chaos, or universal nebula, from which all stars supposedly originated. Further, in the explanation of how this Chaos gave rise to stars, it is supposed that centers of attraction arose at the places of maximum density.

But, since in the original Chaos the density was assumed to be uniform throughout, the question naturally arises as to what miracle could have given the start by condensing some spots and rarefying others. In fact, this theory leads us back more obviously to some creative miracle than even the second law of thermodynamics.

In the last half of the nineteenth century further facts about the stars and about the solar system in particular began to be discovered, which made the original Nebular Hypothesis very improbable indeed, and which necessitated the formation of a new cosmogonic hypothesis. This gradually took shape in the form of what is now known as the Planetesimal Hypothesis, which, though partly based on the old Nebular Hypothesis, has altered the main ideas.

This Planetesimal Hypothesis originated from the theory of tidal friction, as developed by George Darwin. According to this theory, when two dark stars (of which the universe is supposed to be full) come close together, being led to pass close to each other by their respective proper motions, the extreme proximity results in the two dark stars mutually raising immense tides on one another, the tidal friction being so great as to heat both stars to an immense heat, and at the same time to produce in each star a rotation in the plane of the relative velocity of the stars. Incidentally, the mutual attraction of the two stars would probably make a great change in the proper motions of both.

The tidal force acting on each star would furthermore be strong enough to overcome the cohesion of the parts of the star, and thus tear almost all the exterior parts of the star away from the star into the surrounding space, forming a spiral nebula. This nebula, as thus formed, will contain many condensations of larger or smaller size; these, on cooling, absorb surrounding portions of the nebula, and become planets, satellites, asteroids, and meteorites. All these will tend to revolve around their sun in the same direction as their sun itself rotates, as well as to rotate on their own axes in the same direction.

The Planetesimal Hypothesis thus tends to assume that the universe always was somewhat as it is now, but that stars come and go in generations, as it were. There are thus supposed to be at present stars of all sorts of ages; stars of the older generation, and young, warm stars of the newer generation. The very youngest stars are surrounded with nebulas, which are, indeed, usually found to take form of spiral nebulas. The life of a star is somewhat as follows: After the process that has just been described, then first the planets and then the star itself cools off, possibly becoming dark, till proximity with another star comes about again, when the systems are once more heated up and proximity of the two stars generates in both a new planetary system. From the proximity of two stars there issues two new stars. And so the process keeps on from one star-generation to the next.

This Planetesimal Hypothesis is undoubtedly a plausible one, though, like every other theory, there are plenty of observed phenomena that it either does not explain or explains only imperfectly. No doubt, if two dark stars come into such proximity, that proximity will generate new stars with a planetary system to each, as that hypothesis assumes. And it is also true that almost all stages of growth of a star under the Planetesimal Hypothesis are actually observed in the sky—but, as in the case of the Nebular Hypothesis, the initial process is a missing link.

The main co mm on ground of these two hypotheses is, that there was a nebula which condensed into a stellar system, the stars, planets, etc., of the stellar system becoming constantly cooler as they radiate their heat into outside space. So much can almost be observed directly, for all these stages, from the nebula on, are exemplified in the sky. The "ring stage" as supposed in the original Nebular Hypothesis has, however, never been observed in any star, and the only possible example of that is Saturn's rings, which, however, is a spurious example, since spectroscopic observations show that the rings around Saturn are not true rings, but simply collections of small satellites at approximately the same distance from the planet, and which, from this distance, look like rings.

However, practically all stages of the Planetesimal Hypothesis can be seen exemplified in the sky. The Planetesimal Hypothesis does not deny that sometimes a new body can arise by rotation; but, even in such a case, there is no ring process. In the case of the earth, for instance, it is supposed that the earth was rotating with extreme rapidity, the centrifugal force finally elongating it into a sort of pear-shape, the elongation continuing until the centrifugal force at the end of the pear exceeds the gravitational attraction. Then the part of the earth at the smaller end of the "pear" separated and became the moon. Tidal friction afterward slowed the rotation of both parts and the reaction moved the moon away to its present distance from the earth.

The same origin may be supposed for many multiple stars. In fact, the various stages of this process can actually be seem among the stars; for there are variable stars whose variation in brightness could only be explained by this pear-shape, and again there are those whose variation indicates that they are very close "binary" (physically double) stars, and again we have the visible binaries. It may be interesting to note that this process of origin of new bodies by rotation has a remarkable resemblance to the processes of cellular reproduction, only in the latter case surface tension and not centrifugal force produces the constriction and division.

As we have stated before, one strong test of these hypotheses is the observation in the sky of stars in various grades of formation from the nebula to the dark star. The dark star itself is of course invisible, but all grades up to that can be observed, and, in fact, all grades of star development assumed by the Planetesimal Hypothesis, as far back as the spiral nebula stage, are actually observed in the sky. However, the crux of differentiation between the hypotheses is: What was the pre-nebular stage, what brought about these nebulas? The observational method of answering this would be: Do we see in the sky any phenomena that would be likely to lead up to the formation of a nebula?

The only such phenomena that can be observed are the so-called "temporary stars", or Novae. These are stars that suddenly flare up, last a few months, and then gradually fade out. Before the flare-up, nothing whatever is visible in the place where afterwards the temporary star appears; after the star fades out, it has simply radiated a great proportion of its newly acquired light and settles down to the usual brightness of the stars in its vicinity. However, it is a general rule that most temporary stars, if not all, are surrounded by a nebula. We may suppose that the explanation of the nebular condition as generally observed (most nebulas having one or more stars or star-like condensations in the center) could be found in the flare-up of the temporary star.

This explanation would be very satisfactory if we only knew just what happens when a temporary star suddenly appears. It would seem that, by observation, the history of a stellar system cannot be traced back farther than the appearance of a temporary star; so that, if we wish to trace back the development of such a system, it would be important to find out just what makes a temporary star flare up. It seems to be the general consensus of opinion among astronomers that there is nothing in the appearance of a temporary star to make it even remotely possible to assume that they are due to collisions of bodies. A co mm on theory is, that the flare-up is due to explosions of hydrogen. This may sound reasonable, until we notice that hydrogen is not an explosive substance unless in contact with a sufficiently large amount of some such substance as oxygen. We would thus have to suppose a body consisting of hydrogen meeting an oxygen shoal and then exploding. A temporary star, however, consists mostly of hydrogen, and hardly contains enough oxygen to make all that hydrogen explode. Furthermore, at such a heat as that of ordinary stars, still more so at that of temporary stars, water, which is the product of an oxy-hydrogen explosion, could not exist, and its decomposition under the influence of the explosive heat would absorb just as much heat as the explosion produced, thus leaving our temporary star without any heat or light at all.

Under the Planetesimal Hypothesis it has been suggested that a temporary star actually consists of two stars approaching in proximity to each other and drawing out of one another heat and a nebula. This sounds very plausible, but is a bit difficult to support. Besides, it is difficult to see why most of these phenomena should occur in the Milky Way, that is, near the edge of the Flerschel drum. In fact, it is not easy to understand exactly what does happen when a temporary star appears. We may possibly, however, benefit by the more detailed observations taken of Nova Persei 1902, a temporary star which appeared in August 1902, and in connections with which many strange phenomena were observed. This star was first discovered by a man who, though not a regular astronomer, was a habitual star-gazer. One evening in August, 1902, he noticed in the constellation of Perseus a new second magnitude star that he had never seen before. This discovery being made public, it turned out that, on the previous night, a photographic plate of that part of the sky had been taken at the observatory, showing stars down to the twelfth magnitude, and yet the spot where this new bright star appeared was vacant on those plates. Evidently, within 24 hours, the star had flashed up to the second magnitude from a magnitude certainly less that the twelfth, if indeed it gave any light at all; that is to say, it flared up suddenly at least 10,000 times its original brightness, if, indeed, it gave any light at all before the flare-up. One characteristic of this flare-up, then, was its suddenness: the time the star took to flare up in this manner is not known, but it certainly was only a matter of hours.

But a far more interesting aspect of the affair appeared later. The star, indeed, appeared a bit hazy; but soon it was seen surrounded by a nebula, which kept on increasing in size. The nebula was approximately circular in shape, the radius of the circle increasing by about 5 seconds of arc each month, which would make in a year about 1 minute of arc. Since the star showed no parallax, so that its actual distance was too great to be measured, that meant that its distance was more than merely hundreds of light years, but rather ran into the thousands. The rate at which this "nebula in motion" was spreading, being about a minute of arc in a year, must be, in a year, about 1/3400 of the distance of the star (a minute of arch being about that fraction of the radius of the circle). This meant, if the distance of the star was to be measured in thousands if light years, that the rate of spread of the nebula was at least one-third the velocity of light, if not more. The most probable hypothesis was, that the rate was exactly the velocity of light, making the distance of the star about 3400 light years.

Now, since we could hardly suppose that any explosion, however violent it may be, or especially any result of tidal disruption, would produce matter which would actually move in all directions with a velocity so great as that of light, the observers were led to the hypothesis that the nebula was actually there before the star flared up, and that the apparent spread of the nebula was an illusion due to the actual spread of light through the nebula, first in the central parts, then gradually toward the edges. In other words, the conclusion was arrived at that the star was in a nebulous condition long before it began to give out light.

This is hardly in accord with either the Nebular or Planetesimal Hypothesis, for, on the first, light would not be a sudden development, and, on the second, both light and nebula originate at the same time, the light reaching outside points long before the nebula. In fact, we may say as a matter of observation that only the stars which appear to be of the older generation are surrounded by nebulas. We should therefore conclude that a nebula is some super vaporous phenomenon which is only possible as the result of such extreme heat that the vibration of many particles gets them almost altogether away from the influence of gravitation. Thus, the conclusion that the nebula existed previously to the flare-up can only mean that the star was, before it suddenly flared up, in as hot a condition as afterwards. This can only mean that the flare-up could not have been due to the sudden accession of heat that might be supposed under the influence of either tidal friction or of a collision or explosion. The heat was there before, but somehow or other it did not transmit itself into outside space. But, since such transmission of heat into outside space in the form of radiant energy, and in particular, of such great heat in the form of white light, is a consequence of the second law of thermodynamics, and must be a result if the second law of thermodynamics is supposed true, we must suppose that Nova Persei 1902 had all the necessary heat, but that, until that day, the second law o f thermodynamics w as, for some reason, not operative on it.

Before the flare-up, then, the star in question was in a condition in which it showed little or none of the positive tendency. The tendency could hardly have been the neutral tendency, for, as we have seen, the neutral tendency does not form bodies at all, though there may possibly be such a thing as a body going through the neutral stage temporarily, when it is half positive and half negative. It follows, then, that the flaring up of this star must have consisted in its changing over from the negative to the positive tendency. And we may readily assume that similar circumstances give rise to other phenomena of temporary stars; and, since the temporary star seems to be the phenomenon that precedes the nebula, we may come to the conclusion that the pr e-nebula condition of any stellar system is a stage in which that system follows the negative tendency, followed by a sudden change to the positive tendency accompanied by a gr eat out burst of radiant energy.

If we take 18,000,000 as the approximate number of visible stars, and allow about 9 times as many that are dark or too faint to be seen, and take as an average speed of proper motion of the stars 10 miles per second, then, if we suppose that every star, on entering the Herschel drum, with the dimensions we have supposed, flares up as a result of the change from negative to positive, such flareups should happen, on the average, a little more frequently than once a year. This is indeed the average frequency of the appearance of temporary stars; and it is remarkable that most temporary stars appear to be near the surface of the Herschel drum. Accordingly we may take this as the general explanation of temporary stars.

# XIV THE REVERSIBILITY THEORY OF COSMOGONY

We have seen that, according to the results of observed facts, it seems probable that the pre-nebular stage of a star was a negative, a living condition. Now let us see what would be the theoretical result of supposing our theory of a reversible universe, as far as such results may relate to cosmogony.

We have seen that the structure of the universe, according to the theory of reversibility, is that it consists of irregularly shaped sections, alternately positive and negative. In the positive sections all heated bodies give out radiant energy, according to the second law of thermodynamics. In the negative sections, on the contrary, hot bodies, instead of giving out light or other radiant energy, would tend to absorb it and convert it almost entirely into heat, thus heating themselves up with light received from outside sources. This is in strict accord with the reversal of the second law of thermodynamics.

In the first place, when we examine the changes that take place in time, we may first notice that the structure of the universe probably remains somewhat the same always. The positive and negative sections do probably indeed change their position, but, on the whole such change would consist of a general motion of all the sections alike through space, so that the sections do not move relatively to one another. Further, there may be slight changes in the shape of the various sections.

But by far more important is the motion of the individual stars relative to the various sections. The motion of a star, being uniform motion in a straight line under the law of inertia, (the infinite universe assuring us that there will be no gravitational disturbance unless by accident the star should come very close to another star) will have the result that the star will constantly be crossing from one section of the universe into the next, from a positive section into a negative, and from the negative section into another positive one, and so on ad infinitum.

In a positive section of the universe, the star, which was at first hot and bright, radiates its heat into outside space, and gradually becomes cold and dark. We have already seen (in Chapter X) that, as this cooling process goes on, life gradually extends itself at the expense of the opposite, the positive tendency, until, when the cooling process is well under way, life has absorbed practically all inorganic matter, leaving as non-living matter the organic compounds formed by life, which it builds up into pseudo-living organisms. We may suppose that, when a stellar system crosses over from a positive to a negative section of the universe, there happens this slow process of development, of life growth, changing the star from a positive one to a negative one very gradually.

To trace this process of development farther, we must note that

59

the evolution of living stars and planets consists to a great extent in their absorbing radiant energy from outside space and using it to build up higher heat levels in themselves. The life of these stars and planets depends on their being constantly fed, so to speak, with radiant energy uniformly from all directions; which is something that is not obtainable in the positive section of the universe, where the distribution of light is very irregular. In the negative section, however, we are surrounded by positive sections, and in such a way that the light obtained from them is approximately uniform, so that the negative stars and planets contained therein can be properly fed. The tidal forces produce, under those conditions, not tidal friction, as they would under the positive tendency, but a sort of tidal irritation, speeding up all motions of rotation, etc. These living stars and planets, building up in themselves ever higher levels of heat, finally pass into the molten and then into the vaporous stage, and finally the star develops a nebular stage, this nebula taking a spiral form on account of the rotational motion of the star, ever increasing through the process of tidal irritation. Thus we get to a nebular stage; and the dissociation of atoms that goes on in the last part of the negative stage, when we have great heat, will make the stellar system largely one that is constituted of hydrogen, the smallest atom known.

Meanwhile we might expect that not only the star and its planets, but also a number of small masses on the star, would have life, that is, would follow the negative tendency; besides the existence of a number of pseudo living organisms. These simple living masses would, when in the heated condition, also tend to live by absorbing radiant energy from outside space.

Now, we may suppose that, after the nebular stage has been reached, and the star and all its planets are but more condensed vapors in the nebula, the stellar system in question finally comes toward the end of that part of its path which is in the negative section of the universe. The stellar system, nebula and all, is quickly approaching the boundary surface, with the positive section shining brightly ahead of it. The sudden absorption of an immense amount of light from the front will tend to cause a great, sudden, additional building up of heat, so that we will have an immense amount of heat developed before the boundary surface is finally reached. Star, planets, nebula, and all, are constantly absorbing ever more and more heat; including also the smaller living masses on the stars and planets; with the possible exception of the pseudo-living organisms. All are dependent on the constant accession of radiant energy to sustain their life.

Now, when the system comes near the boundary surface, when it is on that surface or very close to it, the accession of radiant energy suddenly ceases to be uniform in all directions, and, once the boundary-surface is crossed, no light whatever is received from behind because light crosses the boundary surface in only one direction, that from the positive to the negative side. The uniform access of radiant energy that the system has to feed on is suddenly cut off, and the stars and planets can no longer continue to live. The proper supply of radiant energy food being suddenly cut off, the death of the system results, and therefore, after crossing the boundary surface, if not a little before, there occurs in the stars and planets of that system the transition from the negative tendency to the positive. As we have seen before (in Chapter IX), the transition in this direction might be theoretically expected to be a sudden and complete one. Hence, somewhere near this boundary surface, we might expect a sudden reversal of this process due to the death of the system, to its suddenly ceasing to be alive as it was when in the negative section of the universe. And the moment this transformation occurs, the second law of thermodynamics immediately begins to apply, and the heat of the system being at a higher level than that of outside space, would suddenly begin to spread itself at a rapid rate into outside space by a sudden outburst from the star of radiant energy. The nebula, being more scattered, gives out much less light, but has to be lighted up to a great extent by the central star. This will produce the phenomenon of the "nebula in motion" as seen in Nova Persei in 1902. In other cases, the nebula itself will give out enough light to be visible immediately.

However, this reason for the "death" of the star, planets, and nebula, does not by any means apply to the smaller living masses that existed on those bodies. The pseudo-living organisms that formerly existed on those bodies will then, with very little change, become simply inanimate bodies; but the small living bodies, unlike the stars, planets, etc., will suddenly get a new and possibly better supply of their food, radiant energy, from the flare-up of the star. Instead of having to depend on the radiant energy coming in from a distance, there is suddenly opened up for them an immense new supply of light on the star itself, or on the planets. The small living bodies thus begin to feed on the dead bodies of the stars and planets. The death of the stars and planets gives an opportunity for new life to develop as a sort of parasite on the dead bodies. From this survival of life, further life on the planets of that system is descended.

After this sudden flare-up, the light and heat would then proceed to fade out gradually, and the system would continue to evolve along the lines indicated by the Planetesimal Hypothesis, this being the precise reverse of the evolution as it took place before in the negative section of the universe. Finally, the worlds cool off, life extends, and, by the time that the system leaves the positive section, life has again gradually extended so as to take on the large bodies. Now the cycle is complete, and we are back at the original stage.

## XV THE PSEUDO-LIVING ORGANISMS

We have seen that organic structure is likely to be found in either section of the universe in the minority tendency. In the case of the negative section of the universe, where most objects are alive, this minority tendency will be the positive tendency. Thus the organic structures in the negative part of the universe are not living but lifeless beings, though having certain appearances of life. These we have called pseudo-living organisms, which, though in certain respects they appear like living beings, yet their motions are of a passive rather than an active character.

Inasmuch as, on reversal with respect to time, a negative universe becomes a positive one, and the inorganic life that is found in the negative section of the universe corresponds to the ordinary lifeless inorganic bodies that we observe, so we may notice that these pseudo-living organisms are the exact reverse of the living organisms that can be observed in the positive section of the universe.

We have seen that, at ordinary temperatures, these pseudo-living organisms, in order to keep existing, must have a constant metabolic process always going on; this being the exact reverse of the metabolic process going on in living organisms in the positive section. In fact, take any process going on in living organisms, and its exact reverse with respect to time will give us the corresponding process going on in pseudo-living organisms.

Take, for example, the sensitiveness that is characteristic of nearly all life. This will, indeed, be found also in the inorganic life in the negative section of the universe. But the pseudo-living organisms have nothing of the sort. They are not sensitive to causes, but to effects; for a small effect may, in these organisms, be the result of a large cause, as we should expect from the second law of thermodynamics. Thus, while all living substance is sensitive to the past, all lifeless substance is similarly sensitive to the future. This is indicated in ordinary physical objects by the fact that it is easier, where both are unknown, to trace the future than the past. The same will be true of the pseudo-living organisms, which are but complicated physical bodies surrounded by living substance. Such organisms will be organized to be able to feel what is coming, but not what has already happened. As to the past, anything in those organisms that may possibly be called feeling would be absolutely blank.

In the case of living organisms, this feeling, it its most elementary form, consists merely of that irritability which we have already identified with the reversal of the second law of thermodynamics. That is, feeling consists, in its most elementary form, of a stimulus releasing reserve energy and making it available energy or else actually using it. On the contrary, the pseudo-living feeling would be exactly the reverse, turning available energy into a store of reserve energy as effectually as may be. In the more complex living organisms, special organs of feeling are developed, which are of special irritability, organs which are specially efficient in extracting available energy out of reserve energy. In fact, in those special organs of feeling (the nervous system) is concentrated most of the mechanical efficiency with respect to extracting available energy to be used as molar motion. Finally, we have the development of a brain, a central organ in which the reserve energy is stored as a result of special stimuli, and which can use that energy to produce molar motion. In the pseudo-living organism, which is the exact reverse of the living organism, this nervous system and brain would constitute a system of extremely low mechanical efficiency, that is, a system for doing as nearly nothing as possible. It would indeed store up immense amounts of reserve energy, or rather, of partly available energy, which would be almost as good as unavailable.

Thus the nervous system and the brain, which in living organisms is the most active part of the organism, would also be found in the pseudo-living organism, with the same size, shape, position, substance, etc., but would, instead of being extremely active, be the deadest part of an apparently dead organism. And the reason for this obvious, if we will but consider. The physical body, and especially the pseudo-living organism, is sensitive only to the future. If something strikes it, or if any other stimulus is applied to it, this immediately becomes a past phenomenon and the organism can no longer take cognisance of it. But should it ever happen that the body itself produces a visible effect in the manner of motion, sound, heat, etc, the body shows it by its internal condition before the effect is produced, though, as soon as the effect appears, this abnormal condition of the body disappears. The body can feel what is going to happen, not indeed what is going to happen to it, but what is going to happen as a result of it; and the moment the event happens, all is forgotten, as it were, that is, no resulting internal condition is noticeable. In pseudo living organisms, special lifeless organisms built up by living surroundings to resemble in certain respects the living beings that we see, these phenomena will, in the more complex cases, be specialized into a nervous system. Thus the phenomena under the positive tendency, and in particular in the pseudo-living organisms, that are analogous to feeling, refer not to past causes, nor indeed to future causes (this not being the true reverse of past causes), but to the direct reverse of past causes, namely, to future effects.

Where, in a living organism, we have enough complexity to find such an organ as a brain, we immediately have the brain reactions which are known as mental phenomena. These are the centralized stores of available energy which the nervous system has extracted from the outside reserve energy, and which can be used under a stimulus to produce molar motion. The mind i s thus part of the brain-machinery, a highly complex and specialized machinery for the extraction of reserve energy and its final conversion into molar energy. The extraction of reserve energy in the original process is sensation; the energy stored up in the brain at a higher level is the mental process; and this mind can only feel sensations, and retain traces of processes, that have already happened, and refer them to the past. On the contrary, in the pseudo-living organism, the similarly complex and specialized process will merely produce reserve energy for the outside world to use, and any mental process in such organism could only refer not to the past causes but, like all feeling under the positive tendency, to future effects, which, however, would be felt as stimuli and not as effects; for the object itself would be under a strain as if stimulated. In other words, this pseudo-living mind, this machine for doing nothing as effectually as possible, could only perceive and remember the future, and would conceive of that future as the reverse of what it really is, namely, as stimulus instead of effect.

Since the ordinary organic bodies are the simplest form s out of which the pseudo-living organism develops as a high degree of complexity, just as the inorganic life of the negative section of the universe is the simple form of which living organisms are a higher development, we may easily suppose that ordinary inorganic bodies such as we constantly observe have this reversed feeling; but, as mental processes are a result of a highly complicated and specialized organism, we cannot attribute to ordinary physical objects anything like a mind.

It has been a favorite theory of the late Prof. Josiah Royce that physical objects are alive and even endowed with a mind, but that we cannot communicate with them or observe that mind on account of the difference in reaction time. According to his theory, while we react to a stimulus in, let us say, a tenth of a second, let us suppose that there is a being that reacts in a thousand years. The motions of that being will be so slow that to us he will appear practically motionless and dead, while, on the other hand, our motions will be so rapid that he will be totally unable to perceive them, so that he will also think us dead. This theory indicates that a difference in reaction-time might be the cause of our not attributing life and feeling to physical objects. Under our theory of reversibility, the same will be true, only the reaction-time of a physical object will not merely be different from ours, but negative, so that all means of our observing the similarity would be cut off

This does not, of course, mean, that there are no observations or experiments possible from which we could indirectly infer such similarity, but merely that we could not possibly observe it directly, because it is superficially different almost in kind from living feeling. It is not, of course, quite true, that physical objects do not show the effects of stimuli; they do indeed, in some cases, but to a markedly less degree than they show the incubation of future effects. Thus, if this sensitivity could be at all called feeling, a physical object, once an event is past, would feel it vaguely if at all, and with a great uncertainty. To the pseudo-living organism, the past has the same vagueness and uncertainty as the future has for us, though some dim guesses as to the past might conceivably be made by the pseudo-living mind.

But it still remains true, that if we were transported into a negative section of the universe, though the pseudo living organism would appear in shape, substance, structure, etc., exactly like the living organisms we are accustomed to, yet we should not recognize the existence of sensitivity or mental phenomena in them at all, and they should appear to us as lifeless bodies, which indeed they are. They would appear to us merely as extremely well preserved corpses. And, because we cannot feel what the pseudo-living analogue of a mind would conceive as a stimulus, and would not react to it, those organisms would similarly think of us as dead.

## XVI PSYCHOLOGICAL ASPECT OF REVERSAL

This matter brings up the question as to how the pseudo-living analogue of a mind, this "machine for doing nothing," would conceive of its own portion of the universe. In trying to solve this question, we must remember that its memory is directed not toward the past but toward the future; because, memory being but the stored-up feeling in a higher form of development, and feeling being that of reserve energy, it follows that feeling and energy must, in any organism, be directed towards that direction in time in which that organism had less reserve energy, and away from that direction of time in which the organism acquires more available energy.

Now, we know that the method by which we really distinguish between the past and the future is by the fact of our remembering the past, while the future to us is an uncertain matter. It follows, therefore, that to the pseudo living mind, the past will be conceived of as future, and the future as past. An organism conceives, therefore, of the flow of time, in the inverse direction to that in which its memory is directed, that is, in the direction of time in which that organism builds up reserve energy into available energy. Or, since organic phenomena are found in the minority tendency of a given section of the universe, such an organism must conceive of time as flowing in that direction in which the majority tendency, that is, the general surrounding world, decreases the amount of available energy and increases the amount of reserve energy. In other words, an organic being, whether living or pseudo-living, must conceive of time as flowing in such a direction that the second law of thermodynamics prevails, independently of whether that conclusion is correct or not.

This arises from the fact that the pseudo-living organism, though existing in a world in which the second law of thermodynamics is regularly reversed, does not perceive its surroundings as they are, but, on account of the fact that it is not life, but reversed life, it perceives the world as reversed in time, its perceptions form a sort of time-mirror, which would thus produce the illusion of reversal, with the result that such a perception would show the organism itself as alive (not as a pseudo-living organism) and the surrounding world, which is really alive, as lifeless and as following the second law of thermodynamics.

Thus, if we were pseudo-living organisms in a universe the exact reverse of ours, that is to say, in the corresponding part of the reverse universe, we should, as pseudo-living organisms, be under this reversal delusion and conceive of the world and of ourselves exactly as we do now, and, in fact, we would have exactly the same ideas as now in relation to everything. Thus there is actually no way for us to tell whether we are living organisms in a positive universe or pseudo-living organisms in a negative universe; in both cases, the former would be the apparent situation. Under the conditions under which a complex organization like a mind can be produced, that mind must conceive of its surroundings in such a way that the second law of thermodynamics would follow. It may be that the law is or is not a physical fact in that particular part of the universe, but conceiving of things in that manner is a necessity for an organized mind. In other words, the second law of thermodynamics i s not a physical but a mental law.

However, this must be construed with limitations. There are certain physical facts as to whether the second law of thermodynamics is actually true or not in any given part of the universe. We cannot say that the real universe and the reverse universe are one and the same on the strength of this reasoning; for were we transported into the reverse universe, we should notice the difference; and similarly the pseudo-living organisms transported into our real universe would also easily perceive the difference. But the difference rather suggests the difference between right and left rather than anything else. There are many substances which form two species, one with right-handed molecules and one with left-handed molecules. The reaction of two right-handed substances is the same as that of two similar left-handed substances; but we get entirely different reactions if a right-handed substance of one kind is brought into contact with a left-handed substance of the other kind.

We must regard similarly the difference between any possible combination of events and its reverse with respect to time. They are similar, and at the same time different, in much the same manner as right and left. There is no really essential difference between the forward and backward direction in time, any more than the difference between right and left is an essential one. Time is really a two direction phenomenon, and the two directions are practically interchangeable, instead of being a single direction flow with one direction essentially different from the other. The fact that the two directions of time appear essentially different is due to the fact that our mind is so constructed as to face one direction. There might seem to us to be an essential difference in space between the forward and backward directions, if not for the fact that we are able to turn around.

The pseudo-living mind is one in all aspects like ours, with the difference that it is so constructed as to face the other direction in time; and it has the illusions of difference between the two directions accordingly. To any mind, the past is merely the direction of time which the memory faces, and the future is the opposite direction of time. Hence the pseudo-living mind will see past where we see future, and vice versa. "The first shall be the last, and the last shall be the first"—for the pseudo-living mind. And the reason that there is no way of telling whether we are living organisms in a positive universe, or pseudo living organisms in a negative universe, is that the difference is really one merely between the two directions of time, and, though those two directions are opposite to each other, they have no physical properties which are in any way different.

There are other cases of such conjugate relations, where two phenomena are different, but can be mutually interchanged without the possibility of any test to indicate the difference. The case that is nearest to that which we are considering, is that of any two opposite directions in space. If two opposite directions in space were interchanged, we should merely have a mirror world, but no different physical properties; and, if we were to suppose that, in that world, we should conceive of right as left and vice versa, there would be no way to tell such a world from the one we live in.

A much more clear-cut case of such a conjugate relation is to be found in the domain of algebra, when dealing with imaginary quantities. The quantity i is defined as the square root of -1, but we might remember that any quantity except zero has two square roots, each the negative of the other, so it is with -1; and we thus get two quantities, i and -i. Now, it makes absolutely no difference to any possible formula in connection with the theory of the imaginaries, which of the quantities we call i, and which -i; they are absolutely interchangeable; and yet the two quantities are anything but identical. For instance, the difference between two identical quantities is zero; and yet the difference between these two quantities is anything but zero, but is twice one of the quantities. The difference can be made to be twice either of the quantities, according to which is subtracted from the other.

In fact, we may notice that perfect interchangeability is not identity. The test of identity is, not that the two things may be interchanged in any statement without vitiating the truth of the statement, but rather that either may be substituted for the other in any statement without vitiating the truth of the statement. In applying this test for identity of A and B, we should substitute A for B without at the same time substituting B for A.

We may then say that the mind conceives of time as flowing, because the mind is not symmetrical with respect to the two directions; it faces one direction, according to the laws governing the special machines that would have to pump reserve energy, and therefore according to the phenomena manifested by reserve energy; and, under the conditions which produce such mechanisms, the resulting law is that an organized mind must conceive of time as flowing towards that direction in which is more reserve energy in that particular part of the universe. This may be either direction in time, either that which is, in our particular minds, forwards or backwards; but, if we conceive of past and future with this mental definition, the second law of thermodynamics follows as a necessary mental law. True, were we transported to a negative section of the universe, we should not see things as conforming to the second law of thermodynamics; but the chances are very small that we would be able to live under those special circumstances, under which a sensitive, living air might take an aversion to our breathing it, or, what would be more likely, would send us its carbon dioxide and leave the oxygen for itself, as it would do to the pseudo-living organisms.

If we represent the percentage of "available energy" in a given part of the universe by a curve showing the variation of that percentage through time, we get a wavy curve, resembling somewhat the sinusoid. If the past is placed at the left, and the future at the right, then, as we go along the curve from left to right, the upward sections of the curve represent the negative portions, and the downward sections the positive portions. Time, then, is a two-dimension affair, like the bottom axial line; but a mind in any part would conceive of that time as a flow towards the lower part of the curve, though that may actually take it towards the past instead of towards the future. To that mind, however, no difference is noticeable.

In the diagram the abscissa represents time, and the ordinates the percentage of available energy in the particular section of the universe. The law then is, that whatever kind of mind would be produced under the various circumstances would be so constructed as to conceive of time as flowing towards the lower part of the curve, that is, towards the troughs of the waves in the diagram; while memory would always look towards the crests of those waves. It makes no difference whether either of those directions is actual past or future, that is, on the diagram, whether these directions point towards the left or towards the right (left, on the diagram, representing past, and the right representing future). In

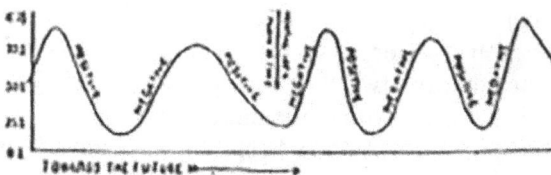

relation to the physical time, the second law of thermodynamics may or may not be true; but, as far as concerns the mental conception of time, the second law of thermodynamics must be true as a majority tendency in that particular section of the universe. Hence, in the last analysis, the second law of thermodynamics is to be interpreted as a mental law, as the law determining the direction in which a given mind will conceive of time as flowing.

It must further be remembered that time itself is not a mental phenomenon, but only the appearance of flow. There is actually no more flow in time than in space, and either direction in time may be called past and the other future, without any difference in the properties of the universe. But the actual existence of intervals of time we must assume as being a physical reality, and absolutely necessary for the explanation of physical phenomena.

Inasmuch as it makes no difference in which direction we suppose time to be running, and we may fix either direction arbitrarily without changing the physical properties of the universe, it is more convenient, in order to avoid any dispute as to the nature and direction of time, to call that direction past in which our memory points, and to call that direction future towards which we conceive time as flowing. In relation to this direction of time, then, we may say that our own section of the universe is positive, and that in that section the second law of thermodynamics prevails.

In fact, we may readily conceive of time as a sort of fourth dimension of the universe. This could readily be done theoretically, only there is a different relation to physical objects. If we used such a conception, we should have to imagine each particle as a sort of thread infinitely extended in the time-dimension. And, further, measurements in time cannot be compared with space measurements. But, although we should not suppose that what we have is a network of threads in a fourdimensional space, yet we can use this as a possible illustration to show what a two-direction time is.

Suppose, then, a four-dimensional space with a perfectly stationary loom, full of threads entangled in all sorts of ways. The ends of the loom we must suppose to be removed to infinity in their respective directions. If, then, we suppose a three-dimensional film to be moved downwards through this loom, the cross-sections of threads would change about so as to appear as the motion of particles. If, now, we suppose that certain sections of thread have some sort of consciousness, and can perceive what is in the film when the film passes them, and their previous condition (or, in other words, the conscious section just above that part), we should have the effect of mental activity. If instead of supposing this film, we now simply suppose that certain sections of thread are conscious, and that each cross-section can perceive only the surrounding objects of its own level and the higher cross-sections of itself, then we may say that each cross-section can perceive the higher cross-section, but not vice versa. This would give the impression of a flow from the higher to the lower cross-sections, thus giving the illusion of one flowing and three stationary dimensions; in other words, of one dimension of time and three of space. Probably this is not the correct explanation of the conception of time, but it illustrates the fact that the two opposite directions in time are no more different than two opposite directions in space.

If we suppose, in the illustration, that any conscious bunch of thread always perceives parallel cross-sections in the direction in which the threads are less entangled, it will give the illusion of flow of this fourth dimension, but in such a direction that motion of particles will always seem to scatter. That is, if the threads in this illustration are constituted to perceive in that manner, they will not merely conceive of one dimension as being time instead of space, but they will actually conceive of that time as so flowing that the second law of thermodynamics will be true.

Though all this is but an illustration, we may conclude: The second law of thermodynamics is really a mental law indicating the direction of the illusory flow of time. Time itself really exists as a two-direction affair, and really has no more flow than space.

## XVII GENERAL SUMMARY OF THE THEORY

According to our theory of the reversibility of the universe, the second law of thermodynamics represents one of two opposite tendencies found in the universe in equal proportions. These tendencies we have named the positive and the negative tendency. The positive tendency is that which follows the second law of thermodynamics, while the negative tendency reverses it. The phenomena of the two tendencies correspond to each other to the smallest detail, each being the reverse of the other with respect to the time-element. Thus, a moving picture of the negative phenomena could be obtained by taking a moving picture of ordinary, that is, positive, phenomena, and running the reel backwards when the reel is being projected onto the screen.

The ordinary physical bodies obey the second law of thermodynamics, that is, they belong to the positive tendency; while living bodies, on the contrary, follow the negative tendency, and therefore reverse the second law of thermodynamics. If we reverse ordinary events with respect to time, as, for instance, with the device of running a motion-picture reel backwards, the living and the lifeless would change places, though, indeed, the shapes and the structures of everything would remain unchanged. So, also, would every physical law not derived from the second law of thermodynamics, so that everything in such a reversal could be explained on the basis of the ordinary physical laws. The reverse of the ordinary physical body is an organic form of life; while the reverse of an ordinary living body is what we have called a pseudo-living organism, having the organic structure of life but not its vital activity.

Occasionally, in moving pictures, in order to get an effect which cannot be obtained in actuality, such as a man going up a smooth vertical wall, the device of reversing the reel is used. In watching the picture produced by such a reversed reel, an apparently unnatural effect is noticed, though it is difficult to say what is so unusual about it. For instance, in one case, a motion picture represented a number of persons diving into the ocean from a high springboard and finding under the water something that frightened them. They were then represented as immediately jumping backwards out of the water on to the springboard. This last part of the film was obviously a reversal of the part representing the diving; but it was noticeable that there were circular water waves converging towards a center before anyone came to the surface, and just as the waves came to the center to produce a big splash, the undercurrents brought the people to the surface while, instead of jumping, the picture represented them as being splashed by the water into the air. The people themselves, on the other hand, lost in this reversal all appearance of activity; around them the water and everything else was jumping and moving, they were being moved in a passive way, as though the water and springboard were living and they dead.

Another way of expressing the distinction between the two tendencies is by drawing the distinction between available energy, energy which can be used under the second law of thermodynamics, on the one hand, and reserve energy, energy below the level required by that law, on the other hand. Of the energy of the universe, part comes under one heading and part under the other. The positive tendency uses up available energy and builds it up into a store of reserve energy, while the negative tendency, on the contrary, utilizes that store of reserve energy that the positive tendency has built up and creates available energy out of it once more. In other words, lifeless objects build up the energy of the universe into a reserve store, which they themselves cannot use; for them, the energy is running down into an unavailable form. But there are always present living bodies which utilize the reserve energy and again build it up into an available form.

Our section of the universe is one in which the positive tendency prevails; but this is true for a finite section of space; in general, there are certain places and times in which one tendency prevails. Taking a given amount of time, this division between the two tendencies divides space into an infinite number of approximately brick shaped sections, alternatively positive and negative. When we are in a positive section, we can see only the particular section we are in, though we may have other evidence (e.g., gravitational) of matter beyond that section. A stellar system, as it moves from one section into another, gradually evolves from a set of lifeless bodies with life on them, through a living stage where there are some pseudo living organisms, into a nebular stage, then finally, on entering a positive section, becoming a "temporary star" and going through the reverse process, from the nebula back to the cooler stages.

One tendency is as universal as its opposite. Life must be found everywhere, under all conditions, precisely as lifeless bodies are. There is no spontaneous generation of life, and therefore life can be traced back as far as we can trace back the matter of which the solar system is made, that is, to an eternity past.

But the basis of the distinction is that living bodies are sensitive towards the past, and lifeless bodies are sensitive only towards the future. If a lifeless body can develop a sufficiently complicated organic structure to manifest mental phenomena, or anything analogous, this sensitiveness towards the future involves a memory of the future only, and, as a result, an illusion of a flow of time from the future towards the past, instead of the reverse as we suppose it to be. The sensitiveness of living bodies toward the past and of lifeless bodies toward the future is due to the fact that, under the second law of thermodynamics, large causes are likely to produce small effects, while, under the reversal of that law, it is small causes that are likely to produce larger effects. Another consequence of the same fact is, that lifeless phenomena are more easily explained by their causes, while living phenomena, on the contrary, though equally the rigid result of causality, are to be more easily explained by the future chains of the causal relation, or as that which is to produce certain effects. That is, living phenomena, phenomena which follow the negative tendency, are characterized by an apparent teleology or functionality that is absent (or at least, apparently so) in lifeless phenomena.

There are also the properties of both tendencies as majority or as minority tendencies. For instance, in one part of the universe, the positive tendency is a majority tendency, and the negative tendency is a minority tendency. In other parts of the universe, on the contrary, the reverse is the case: the majority tendency is the negative, or life, while the lifeless phenomena constitute the minority tendency. We may note that there are various characteristics of the minority tendency such as the formation of complex endothermic compounds and of an organic structure; while the majority tendency, whether positive or negative, is characterized by an inorganic structure and the formation of exothermic compounds. The minority tendency, again, whether positive or negative, is characterized by a metabolic process wherever there is not too much heat to permit of such chemical reactions going on.

We may use the diagram in Chapter XVI to illustrate the alternation in any part of the universe between the positive and negative tendency, remembering that the lower parts of the curve represent a condition where there is less available and more reserve energy. We may make an additional remark on the curve in that diagram, that atoms integrate where the curve is concave towards the left, and dissociate where the curve is concave towards the right; in other words, the concave side of the curve always faces that direction in time toward which we find smaller atoms.

Besides the positive and negative tendency, there is also a bordering tendency, which we have called the neutral tendency. This is comparatively rare, and it suffices to say that it has no tendency even to form compound particles, but remains decomposed into the separate ultimate particles. It therefore is not to be found (unless maybe for a single moment in time) in any known substance, for no substances will be formed under the neutral tendency. But the neutral tendency is probably to be found in the spaces between the heavenly bodies, where it represents the phenomenon of a substance with impenetrability but with no resistance to the passage of a body through it.

We may tabulate as follows the similarities and differences between the positive and the negative tendency:

| THE POSITIVE TENDENCY | THE NEGATIVE TENDENCY |
| --- | --- |
| 1. Follows the second law of thermodynamics. | 1. Reverses the second law of thermodynamics. |
| 2. Decreases difference of energy level. | 2. Increases difference of energy level. |
| 3. Forms unavailable reserve energy. | 3. Uses this reserve energy. |
| 4. Uses up available energy. | 4. Forms available energy. |
| 5. Lifeless; appears passive. | 5. Living; appears active. |
| 6. Inelastic collisions. | 6. Super-elastic collisions. |
| 7. Mechanical efficiency less than 100%. | 7. Mechanical efficiency over 100%. |
| 8. Larger causes produce smaller effects. | 8. Smaller causes produce larger effects. |

| 9. Explained easiest by cause; apparent rigidity of causality. | 9. Explained easiest by effect; apparent teleology. |
|---|---|
| 10. Appear living when reversed. | 10. Appear lifeless when reversed. |
| 11. Absence of irritability. | 11. Irritability |
| 12. Atoms integrate at great heat, otherwise dissociate. | 12. Atoms dissociate at great heat, otherwise integrate. |
| 13. Chemical reactions tend towards exothermic compounds. | 13. Chemical reactions tend towards endothermic compounds |
| 14. Spontaneous and complete generation from opposite tendency possible. | 14. Spontaneous or complete generation from opposite tendency impossible. |
| 15. Can generate opposite tendency only by gradual growth from a living center. | 15. Generates opposite tendency, spontaneously, suddenly, and completely. |
| 16. Partly remains when there is transformation into the opposite tendency. | 16. Is needed if more is to be formed. |
| 17. Hot bodies give out light etc. | 17. Hot bodies absorb light etc. |

| 18. Light tends not to enter the positive section of the universe. | 18. Light tends not to leave the negative section of the universe. |
| --- | --- |

## AS A MAJORITY TENDENCY

| 19. Tends to include exothermic compounds. | 19. Tends to include exothermic compounds. |
| --- | --- |
| 20. Ordinary lifeless objects. | 20. Inorganic life. |

## AS A MINORITY TENDENCY

| 21. Tends to include complex endothermic compounds. | 21. Tends to include complex endothermic compounds. |
| --- | --- |
| 22. Pseudo-living organisms. | 22. Living organisms. |
| 23. Metabolism. | 23. Metabolism. |

## OTHER MISCELLANEOUS PROPERTIES

| 24. Obeys the three laws of motion and the law of gravitation. | 24. Obeys the three laws of motion and the law of gravitation. |
| --- | --- |
| 25. Conservation of mass and of energy. | 25. Conservation of mass and of energy. |

| 26. Sensitive only to the future. | 26. Sensitive only to the past. |
| 27. Organisms conceive of time and events as reversed. | 27. Organisms conceive of time and events in the order in which they occur. |
| 28. Memory must refer to future. | 28. Memory must refer to past. |
| 29. Illusion in positive mental phenomena of flow of time from future to past. | 29. Illusion in negative mental phenomena of flow of time from past to future. |
| 30. All positive phenomena fully determined by either cause or effect. | 30. All negative phenomena fully determined by either cause or effect. |

## XVIII SOME OBJECTIONS TO THE REVERSIBILITY THEORY

The reversibility theory of the universe, which has here been set forth, is only advanced as a mere speculation, as a possible hypothesis. There are many objections to the theory of the reversibility of the universe. All that is intended here is, not to prove this theory scientifically, or even to claim it as perfectly consistent with itself or with facts, but merely to indicate that there are, on the question of reversibility, other possible theories than the one at present generally accepted by physicists, and yet not more absurd or more inconsistent with facts. The theory of the second law of thermodynamics, as we have seen, leads to many absurdities, and we have seen a number of facts indicating the possibility of a reversal of the second law of thermodynamics. Having examined all the facts and all the arguments that we have already marshalled on the side of this theory of the reversibility of the universe, we might as well examine a few facts mitigating against that theory in favor of the generally accepted theory of the universality of the second law of thermodynamics.

In the first place, one essential point of the reversibility theory is the supposition that there are such things as reversals of the second law of thermodynamics, and that those reversals, in our section of the universe, constitute the phenomena known as life. This immediately involves the question: Does life really reverse the second law of thermodynamics? We have no proof that it does, and it may indeed be considered extremely doubtful whether it does. If we look down the list of properties of the negative tendency, both in general and as the minority tendency, (excluding the name "life," wherever it occurs in that list), we will still find that nothing under our direct observation except life could come under that heading, so that, unless life is that reversal, there are certainly no reversals of the second law of thermodynamics within our observation. Now, among the characteristics of these reversals we find "super-elastic collisions" and "mechanical efficiency over 100%." These are very important distinguishing characteristics, and yet it is doubtful if we could find a single authenticated instance of a super-elastic collision occurring even in the case of living beings; and, if that is possible, it certainly is true that inelastic collisions are the more common phenomena with living as well as with lifeless bodies; which would not be true if life were a reversal of the second law of thermodynamics. For instance, when we clap our hands together, it does not result in a larger rebound; in fact, the rebound is slight, and the applause, if kept up a long time, produces a distinct sensation of heat, the heat finally subsiding. Here we have an inelastic collision, molar motion resulting in heat, which finally runs down to a common level—exactly what we might expect under the second law of thermodynamics. In other words, there certainly are living phenomena which obey the second law of thermodynamics instead of reversing it.

Furthermore, turning to chemical products, the final product of living forces is carbon dioxide, the most exothermic compound of carbon. Does not this, then, indicate a process of production of carbon dioxide by the oxidation of more complex or more endothermic carbon compounds, precisely as we might expect under the second law of thermodynamics, and precisely as we should not expect under its reversal? And the accompanying result is the liberation of a large amount of chemical energy, which might be used to explain the energy of life without having recourse to the reserve energy which the second law of thermodynamics says is unavailable. In fact, this process of the oxidation of carbon compounds to carbon dioxide is a process distinctly characteristic of the positive rather than of the negative tendency, thus indicating that life obeys the second law of thermodynamics instead of reversing it.

In other words, it would seem as though our theory of the reversibility of the universe is based merely on a superficial resemblance of living phenomena to the theoretical appearance of a supposed "negative tendency." The actual proof that such a "negative tendency" is actually to be found is wanting and it seems, indeed, extremely doubtful whether it does exist.

Further, in our theory of temporary stars as stars which are already in a heated condition, and with a nebula, but suddenly give out light on taking on the positive instead of the negative tendency, we may notice that we have depended on the peculiar observations of the "Nebula in Motion" in the case of Nova Persei 1902. It certainly is remarkable that, while temporary stars are seen about once a year or oftener, that such phenomena should leave been observed on only this one star. Thus a theory of the universe and of the evolution of stellar systems based on the observations of this star is one that is not likely to apply in general, because this star is an exceptional phenomenon. If the reversibility theory is correct, the phenomenon of the Nebula in Motion should be much more common than it really is.

Further, the theory of the reversibility of the universe supposes that life exists under all sorts of circumstances, even on such hot bodies as the sun. Certainly on the sun there is no possibility of anything of the sort that comes under the heading of life within our experience. If life exists on the sun, it must certainly be so different from anything that we are used to call life, that there could hardly be any points of resemblance. And similarly under many other circumstances such as complete absence of air, water, or both, as, for instance, on the moon. In short, as much as we can observe of life would rather seem to indicate that life, far from existing under all sorts of circumstances everywhere in the universe, is rather an extremely complex phenomenon that can only exist under very special circumstances.

Furthermore, if we come to the conclusion that the second law of thermodynamics is fundamentally based on a conceptual illusion, it would be just as logical to admit the same possibility for the other physical laws, in which case the theory of the reversibility of the universe would almost carry with it its own refutation, since the reversible physical laws are the foundation of that theory. If observed facts can be explained in one case as a conceptual illusion, why not in another case?

## XIX CONCLUSION

The reversibility theory is presented here, as I have said before, as a possibility to be considered. It being a mere hypothesis, I have considered it fair to present not only the hypothesis itself and the arguments for it, but also the arguments against the hypothesis. I may also state that I cannot supply any satisfactory answer to most of the objections stated in Chapter XVIII.

Having thus presented a general theory of the phenomena of the universe, I will now leave the reader to compare for himself the reversibility theory with the prevailing theory of the second law of thermodynamics, and weigh the various arguments for and against each, as they have been here presented.

# Afterword by the Editor

Sidis wrote The Animate and the Inanimate to expand on his ideas about the origin of life, cosmology, and the potential reversibility of the second law of thermodynamics via Maxwell's Demon, among other topics. It was published in 1925, but Sidis may have been working on the theory as early as 1916. One motivation for the theory appears to be to explain William James' "reserve energy" theory, which proposed that people subjected to extreme conditions could use "reserve energy." Sidis' own "forced prodigy" upbringing was the result of putting the theory to the test.

This is one of the few works by Sidis that was not written under a pen name. Sidis writes in The Animate and the Inanimate that the universe is infinite and contains sections of "negative tendencies" where the laws of physics are reversed, juxtaposed with "positive tendencies" that swap over time epochs. He claims that there was no "origin of life"; life has always existed and has only evolved. Sidis adopted Eduard Pflüger's cyanogen-based life theory, citing "organic" things like almonds that contain non-killing cyanogen. Almonds are a strange anomaly because cyanogen is normally highly toxic.

Sidis entertains the idea that life originated on Earth from asteroids (as put forth by Lord Kelvin and Hermann von Helmholtz) while describing his theory as a synthesis of the mechanistic and vitalist models of life. Sidis also claims that stars are "alive" and go through an eternally repeating light-dark cycle, with the second law reversing in the dark portion of the cycle. Sidis' theory was dismissed upon release, only to be discovered in an attic in 1979. Buckminster Fuller (a Sidis classmate) wrote to Gerard Piel in response to this discovery: Imagine my surprise and delight when I was handed a xerox of Sidis' 1925 book, in which he clearly predicts the black hole. In fact, I think his entire book, The Animate and the Inanimate, is a great cosmological work. I find him focusing on the same topics that fascinate me and reaching roughly the same conclusions that I have published in SYNERGETICS and will publish in SYNERGETICS Volume II, which has already gone to press. As a Harvard man of a later generation, I hope you are as excited as I am that Sidis went on to do the most magnificent thinking and writing after college.